Human nature has the tendency to avoid the obvious. Problems in science often resist discovery because the understandable is not appropriate to one's immortality formula. Atheism, for example, is a system of denial designed to camouflage one's responsibility toward the evident. Science will endure years of paradox rather than accept a harmonic solution that might give us a peak into the mind of our creator. Avoidance is not really due to the need of proof; rather it is inherently a psychological effort to avoid perplexing questions about God and existence. Science requires the study of knowing, and no one wants to change their minds after years of study in order to give sway to something self evident. The priests in the laboratory are often as dogmatic as many priests in religion. This obstinacy is based upon the inaction they attribute to God and not to observation, experience, and sound meaning. Understanding things meaningful requires the study of reality beyond the physical, yet scientists limit reality to the physical in order to protect certain unconscious formulas that they have manufactured to cover their ignorance of the infinite.

Samuel Dael

The Einstein Illusion is the second in a planed series of books by Samuel Louis Dael that deal with a classical epistemology towards philosophy, science, psychology, economics, religion, and politics. The first version of *The Einstein Illusion* was completed in 1989 as a fundamental text on classical physics, but challenged modern theory by providing a classical alternative to relativity's paradox. Little has changed since then, and much has been added in this version about electromagnetic theory, gravity and updated relativity concepts.

Authors Publications:
The Platonic Idiom   *2008*
The Einstein Illusion   *2010*
The Darwinian Descent   *when God wills*

The Einstein Illusion

Samuel Louis Dael

Vision Impact Publishing
P.O. Box 1338, St. George, Utah 84771
editor@visionimpactpublishing.com
http://www.visionimpactpublishing.com

This book is dedicated to Einstein, and to the Unified Field Theory that has been neglected since Einstein's death. Modern physics has carried the infirmity of the Einstein paradox for over one hundred years. Rather than form a harmonic complimentary principle, modern physics is still in the age of mysticism. This book is not written to those who look to modern physics as a solution, but to those that look prior to Einstein and to Michael Faraday, the neglected prophet of a true physical reality.

Samuel Dael

The Einstein Illusion

Copyright © 2010 Vision Impact Publishing
All rights reserved.
Printed in the United States of America
Samuel Dael http://www.samuel-dael.com

ISBN 978-0-9827313-0-7

Einstein has become a system for many, a monument people have built, a symbol that they need for their own comfort.
                    Leopold Infeld "Albert Einstein The Man"

Contents

Contents vi
Preface viii
Introduction xii
1. A New Reality 1
   The Absence of Faith 1
   A Science for Mysticism 15
   Effect Without Cause 21
2. Striking a Balance 25
   The History of Reality 25
   The Age of Distortion 32
   Reality's Verb 36
3. The Square of Reason 41
   Absolute Basics 41
   Equality In Proportion 47
   The Law of Conservation 60
4. The Geometry of Mathematics 65
   The Subject of Space 65
   Time Has No Direction 70
   All Things From Zero 73
5. Much About Nothing 79
   Breaking the Time Barrier 79
   Classical Illusions 85
   Einstein's Transformations 91
6. The Einstein Illusion 105
   The Doppler Paradox 105
   Electric Contraction 113
   A Bolt of Light 120
7. Objectivity in Matter 125
   Objectively Speaking 125
   Conservation of Mass 132
   Mass-Energy 137

Contents

8. The Predicate Nature of Energy 145
  The Mechanical View 145
  Field Refraction 150
  Quantum Particles 156

9. Relativity and Epistemology 161
  Einstein's Clock 161
  Angular Energy 165
  Solving for Mass 179

10. Action At A Distance 185
  Particle or Field 185
  Field Geometry 191
  A Particle of Intelligence 226

11. Maxwellian Space 241
  What is a Radio Wave? 241
  Radio Antennas 254
  The Nature of Charge 264

12. The Big Illusion 277
  The Primevil Fireball 277
  Friction In The Universe 292
  The Missing Complementarity 297

13. A Matter of Intelligence 313
  The Denial of God 313
  Probability of God 329
  The Tao Of Physics 334

14. Theory Update 345
  Personal Note 345
  Antimatter 347
  Gravity and Dark Matter 351
  String Theory 356

References 361
Index 365

Preface

In recent years cosmologists have continually expressed the idea that our universe is just one universe among an untold number of universes. They use the mathematics from such things as string theory and relativity to suggest the existence of seven extra dimensions if not more. The common reader tends to feel intimidated by modern physics because there is nothing that can be reduced to intelligent understanding. These all-knowing modern priests of cosmology seem to hide behind some exotic equation that can literally never be explained, demonstrated, or even proven. Socratic intelligence demands understanding and cosmologists need to better define their terms. *The Einstein Illusion* by Samuel Dael makes this very attempt to clarify the paradoxes of deliberate complexity. The attempt is to redirect our understanding as if God himself is explaining the elements of creation rather than an equation.

Samuel Dael holds that modern physics has become progressive towards endless forms of mathematical magic where the minds of science claim to be the gods of discovery. They claim this because only they have the intelligence to see the strangeness of the universe that no philosopher could ever imagine. Like witch doctors they favor the mystical, the contradictory, and the elusive that even a God of wisdom could not perform.

Exactly where physics jumps the track of understanding is laid out systematically in *The Einstein Illusion*. Samuel Dael does this by leaning upon ancestral ties to a better discipline of philosophy. It is philosophy that teaches the mind to not accept confusion. It is philosophy that questions and it is science that tests ideas. In philosophy there is no authority but the freedom and responsibility to enquire. In science there is proof taken from observation. But proof from observation does not negate a need for further philosophic enquiry. When we silence the intelligent mind's need to lay hold of clear meaning, we assume a position of infallibility in our observations, no matter how accurate our equations. Put another way, when observation enforces the acceptance of a paradox, the philosophic mind is intimidated to accept it without being able to ask the physicist to explain the workings. For the first time, Samuel Dael explains the workings of relativity because philosophy wants to know; she wants to see why and how. When we ask how we know, this is called epistemology and the starting point of anything philosophic, and yet the modern relativist seems to negate epistemological logic to justify their conclusions. *The Einstein Illusion* is an epistemological discipline the modern physicist has ignored far too long. Mr. Dael puts it better when he says,

> Mass-energy conversion does not imply substance at one moment and non-substance at another. Energy and mass coexists together, but differ as to reality. If the relativist is determined to maintain a conversion and not a relational change from angular to linear momentum, so let him believe in magic.

Mr. Dael realized over thirty years ago when he began his research that if our observations produce a paradox we do not need to change the meaning of words to fit a new observation. This only creates a void for the previous meaning. Take the term warped space for example. At one time space meant empty expanse. Now if this empty expanse is something that can warp, what happened to the original meaning of empty expanse? How can you warp nothingness? Mr. Dael argues that the original meaning of space is still there and the thing that is supposedly warped is not space at all but a field in space.

The same with time dilation, black holes, the expanding universe, zero point energy and many more fabrications in meaning that are essentially changes made to original terms with original meanings tossed aside. According to Mr. Dael, this behavior of redefining established terms is similar to book burning. In *The Einstein Illusion*, Mr. Dael discusses a psychological motive that deliberately seeks to "curve-fit" meaning to mask the real agenda of fear within. And Mr. Dael personally holds modern physics guilty for the sophistic decay that creates new turns of meaning without properly dealing with the classical meanings. The correction Mr. Dael offers is nothing short of a unified field theory holding to both classical meanings of space and time as well as the math in Einstein's equations. This philosophical exercise has produced a model of light with a geometrical explanation that solves for the particle/wave duality. I personally will not be surprised if philosophy herself is not uplifted at the same time by the astounding epistemology offered in this book.

What you are about to read does not fly in the face of Einstein. Instead, it challenges the game of illusionary tricks that modern physicists obtained by the perversion of Einstein's observer-dependent reality. This was done to gain greater self-importance attributed to the relativist's over zealous imagineering.

I must admit that I am a student of the humanities and, while not a physicist, I am an epistemologist. It is with this background that I recommend all of science quickly return to its foundational roots in philosophy if it hopes to advance past the darkness of accepted paradox that has engulfed the modern scholar. I predict *The Einstein Illusion* will launch a redemptive restoration of meaning because there is simply no technological replacement for what used to be a pursuit for the truth. Instead, all that we have is an epistemology of meaning to help discipline the mind to recognize the truth.

Mica Ron Thomas
Editor Vision Impact Publishing

Preface

When a writer seeks any form of truth it is always at odds with the Platonic mind and its progressive tendency to redefine terms used by intuitive intelligence. We often think that truth is discovered by the proper authority. Truth resists discovery and understanding because established authority plays games with reality. Like Plato, the modern theorist simply creates a new argument in order to circumvent a common sense reality and the natural ability of intelligence to perceive. Shuffling reality around has been a constant tendency since Plato. The human mind wants to rewrite reality and destroy the essence of the Socratic mind and the responsibility that comes because of it. Just as Plato used the name of Socrates to alter reality, every modern theorist uses the name of Einstein to distort well understood terms in order to create their own reality formula. They also attempt to nullify the essence of God and of all intuitive intelligence. As with philosophy, the modern theorist is not exempt from the desire to turn reality upside-down.

Samuel Louis Dael

Introduction

Readers and teachers familiar with modern relativity will find *The Einstein Illusion* a version markedly different in both content and meaning. The attempt is: "To make relativity less paradoxical and more self evident." This required a model of light more fitting to Einstein's Photoelectric Effect rather than the traditional wave theories of light.

In finding the abstruse formulations of relativity science incomprehensible, I have fallen back on the meaning of words. The maturation of this approach came from the study of popular and textbook relativity, but mostly from the development of an epistemological philosophy that requires the terms, used in relativity, to hold their meaning with invariant precision.

In other words, "All meaning must be axiomatic." I do not know from where this thought appeared. I do feel that its origin supports the foundation of philosophy. Classical philosophy has always depended on a proper approach of defining the terms including the recognition of self-evident truth based on prior axioms. For this reason I would like to suggest this book to be a classical philosophical approach to relativity—an approach that modern philosophy has failed to provide. I do not mean philosophy as usually practiced today, but in the classical sense, a science investigating the principles of reality. In this sense, this approach not only encompasses physical science, but places faith at the axiomatic center when exploring the relationship of subjective reason and objective truth.

The Einstein Illusion not only provides a corrected definition of light to solve the duality found in modern science but it also injects a long over due predicate reality to compliment the traditional subjective and objective philosophy. This altogether new method places the definitions of words such as conservation, existence, and infinity into invariant relationships rather than relative interaction. Basic words are compared numerically. The meaning of one word has been generalized by the higher and particularized by the lower. One might consider this a Pythagorean method that fixes words into such a rigid criterion that hopefully the meanings are preserved and over-intellectual altering and sophistic decay can be averted.

The customary procedure of thinking follows a process of induction from the particular to the general. In this book, however, I postulate that a better understanding can be achieved by starting from the general and working to the specific and then back to the general. This type of thinking usually does not survive the academic world because the thought processes are slower and do not weather the quick acquiescence of fact. For this reason, the thoughts in this book come from the desert with literally no formal contact with the academic or scientific world.

The effort will be to formulate a proper explanation for relativity without reverting to modern paradox and the closed circular logic found in modern textbooks. A segmental designed reality has become a far better solution. The results produce a sufficiently striking view for the religious scientist, more understanding for the non-specialist and the student of physics, but a trial will certainly surface with traditional science fiction writers of relativity.

Those that contemplate the use of this book should be warned that they will have to abandon the common prejudice that observed fact remains the soundest test of physical understanding. The basis of this conclusion comes from recognizing that observation and fact represent different parts of reality. Too often they are treated as one. Once the traditional objective reality receives the proper action of reason, observation can be reconciled, but not by itself. I strongly encourage the creation of a third middle reality in which to place the act of seeing. Only then can fact stand as objective truth. Observation will then retain power to show agreement between reason and fact and not try to become fact by itself. It should have been concluded long ago that something must be wrong when observation creates paradoxical solutions or when fact and reason contradict. The reason for this is the avoidance of a predicate reality by making the observational verb become the objective fact and forgetting that the object receives the action of reason. Observation only shows agreement—it is not objective fact. It is like objective fact receiving reason through observation. By making observation fact you forget reason and the world becomes a paradox.

I take the position that the unspecialized thinker has access to basic questions which transcend technical proficiency; and that these basic questions can be approachable insofar as one resists getting side tracked by technical muddle, and maintains an awareness of a complimenting three-part whole. This book maintains this whole by explaining how the parts of reality fit together. Modern philosophy has been so busy perfecting reason and objectivity that it has forgotten the philosophy of "apparent action"—appropriately defined in *The Einstein Illusion* as a philosophy of predicativism. I do not, however, mean the highbrow predicativism that has risen since this work was first laid down in the 1980's.

The first three chapters develop this long misunderstood reality by revealing that subjectivism and objectivism do not oppose within the philosophical spectrum, but they are brought into agreement by the use of a predicate reality. Keep in mind that the object receives the action of events, as in observation, and the verb shows agreement between the subjective reason and the objective fact.

When one's parts of speech begin to commit error, the only way of correction demands a return to basics. Just as the verb can never become the subject, and just as the object can never become the verb, so also should reality be set into three parts. The language of mathematics presented in the first chapters suggests that all cannot be one single reality but that space, time and mass exist separately as direction, action and substance.

While never intending to require any computational skill, this thesis contains a little mathematics to promote understanding and not to place relativity into a level of incomprehensibility. The mathematics has been introduced more for the building of a philosophical basis and translating relativity out of its hieroglyphics. In essence, this treaties does not question Einstein's mathematics, it questions the meaning of the terms used in the equations that produce the popular paradoxes. Changing the meaning of the terms creates contradiction. It is not the mathematics.

Once Einstein's Special Theory appears scanty, it can be illustrated epistemologically that the meaning of the terms taken from the Special Theory polluted the General Theory of Relativity. For this reason, chapters eight and nine accept the General Theory as to mathematics and experiment, but not as to epistemology or meaning.

The model of light as a matter field, suggested throughout the book, has been poignantly illustrated in chapter ten. Light curves in matter or matter becomes the geometrical relationship of the magnetic and electric particle fields suggested by Faraday. Light outside of matter would then be the electric particle field as it spirals around magnetic space at rest. Light then would be constant as to angular momentum, but not linear. In this respect, light cannot be energy as to reality. Energy belongs in the predicate, and this would mean the action of matter fields including light. Light then remains the building block of matter. Energy and matter are two different realities. Light then is objective and energy is predicative. Thus conversion is impossible in terms of reality. We have mistaken light to be energy rather than substance that carries energy as any particle would.

It was from Maxwell's equations, depicting Faraday's electric and magnetic particle field, that Einstein developed his equations. If the equations suggest particle fields—why not let light and matter both be different manifestations of Faraday's particle fields?

> The Maxwell equations relate the "partial variations" of, say, the electric field, in space, to the time variation of the magnetic field. They put into precise mathematical formulas just the sort of empirical observation that Faraday made.[1]

Maxwell put Michael Faraday's field relationship into mathematical order, and the epistemology or field geometry illustrated in chapter ten attempts to put Faraday's field description into a geometrical objective reality. The results build a more obvious relationship between light and matter.

Faraday, who was sixty-six at the time, wrote a letter to Maxwell and illustrated a need for which this book was written.

> There is one thing I would be glad to ask you. When a mathematician engaged in investigating physical actions and results has arrived at his conclusions may they not be expressed in common language as fully, clearly, and definitely as in mathematical formula? If so, would it not be a great boon to such as I to express them so? – translating them out of their hieroglyphics, that we also might work upon them by experiment. I think it must be so, because I have always found that you could convey to me a perfectly clear idea of your conclusions, which, though they may give me no full understanding of the steps of your process, give me the results neither above nor below the truth, and so clear in character that I can think and work from them. If this be possible, would it not be a good thing if mathematicians, working on these subjects, were to give us the results in this popular, useful, working state, as well as in that which is their own and proper to them?[2]

Einstein abhorred the conclusions of modern cosmology, but failed to admit that the paradoxes of relativity fostered it. It is basically because the mathematics is not presented in a "useful working state," but rather conclusions are made from improper assumptions about the terms in the equations.

I have introduced a chapter called *Maxwellian Space* since the original work in an attempt to illustrate how equations can be easily distorted as to what actually goes on in order to reach what has been termed 'objective fact'. The outcome is not in question, but the active process in order to reach objectivity is the issue. We too often think equations are a dynamic process rather than a function illustrating points in space and time. Chapter eleven is really a very recent addition since the original work was completed in 1989 and remained unpublished until 2010. I felt that my own original views created certain paradoxes with radio waves that needed to be addressed. In this chapter I attempt to illustrate that the traditional concept of radio and television waves have nothing to do with photons as particles that travel through the immensity of space. Radio and television waves, I attempt to illustrate, are not entirely waves of light or even photons as measured in the photoelectric effect. I expect to be severely criticized for this view, but since I found that physicists and electrical engineers have contradictory explanations about current in a wire, I am inclined to feel justified.

Since the debate as to whether light is a solid particle or a wave comes from differing observational experiments and not contradictory mathematics, and since the previous chapters illustrate that light is neither a solid-like particle nor a wave of pure energy, I upset the modern quantum theory by turning particle probability into field probability. Again, it is not the mathematics that error, it is the epistemological meaning of the terms.

I would find it very interesting for more testing to prove that light not only breaks up during absorption and emission but also upon impact. There are a few experiments of my own that illustrated this in the end of chapter twelve. If all the testing done in the past proves otherwise, I would be surprised and very interested. The above is primarily why chapter twelve attempts a very subtle blow against the big bang cosmology. My case in opposition is general, but logical meaning and not apparent observation builds a generalized form of thinking. To be too particular does not give a panorama of reality. Once it is understood that light is a spiral substance, the premise that the velocity of light does not remain constant in a vacuum everywhere in the universe becomes clear. The Special Theory of Relativity assumes the constant velocity of light, but the general theory proves it to be in error once the meanings of the terms are corrected. This conclusion comes without the aid of traditional observation, but by looking at the many scientific observations differently, with a complimentary understanding of light, the premise is confirmed.

In searching back and trying to find the place where modern science jumped track, I came to a psychologically based conclusion. There appeared to be certain psychological benefits for shifting the meaning of words and thereby changing objective reality. The benefit seemed to cloak one's image of both God and reality. If one finds a wailing and gnashing of teeth, I mean no harm. Relativity, not a reality unto itself, should only have served to test agreement between the subjective reason and the objective fact. If agreement does not appear, something is wrong. Modern science and philosophy could not see it.

Without faith in self-evident axioms, both reason and conservation in existence are altered in order to build a personal elixir of life where all things are relative. Reality reaches a full swing by making it what you want it to be. Why fear? Death is simply all in the mind. The psychology of this is what drives the need to position reality and avoid a three-part process.

Samuel Dael

1. A New Reality

> For life is at the start a chaos in which one is lost. The individual suspects this, but he is frightened at finding himself face to face with this terrible reality, and tries to cover it over with a curtain of fantasy, where everything is clear. It does not worry him that his "ideas" are not true, he uses them as trenches for the defense of his existence, as scarecrows to frighten away reality.
>
> -Jose Ortega y Gasset[1]

The Absence of Faith

Words are bits of language. If properly defined, they form a complete thought and create windows into reality. But when the meaning of a word varies, communication and thought do not provide a clear sense of reality. In other words, the proper relationship of clear and solid-meaning words measures the extent of human knowledge. The origin, nature, and methods in which the meanings of words are designed exemplify the role of epistemology. In this respect, epistemology has always been the language of reality.

Before any strategy begins in an effort to discover the ultimate in objective truth, a wise philosopher will take the trouble to define the terms. Consulting the dictionary for proper meaning may not be a simple task. While each dictionary clearly demonstrates that many words have various meanings—even related definitions and antonyms, simply choosing a definition is not the role of epistemology. Determining the origin of a word in conjunction with its part of speech differentiation exemplifies the true function of meaning. Too often, the thinker selects meaning out of choice or out of tradition, never realizing that the various shades of evolved meaning can be arranged into three connotations. Choosing the right meaning can be just as explicit as choosing the right word. An important example to this argument comes with the word reason. Three connotations will surface as if from three reality points of view.

The definitions of reason are:

1. "Sound mental powers that are concerned statements of logical thinking."

Reason here implies a cognitive (thinking) process—a process where attempts are made to define things free from contradiction. This can be selected as a subjective process and thus a subjective definition.

2. "A basis or circumstance that justifies some action or event."

The action or event may be clear, but the justification may or may not be the real reason. Never the less, the event approach reveals a predicative or active definition.

3. "To form conclusions based on facts."

This meaning attempts to place a conclusion alongside a fact. The fact may be clear, but the conclusion may or may not be correct. Conclusions are by-products of reason rather than part of its process.

We so often fail to look for a common denominator found in improper definitions. Both the second and third above are relative or variant and both are subject to opinion which is nothing but a distortion of subjective reality. In other words, our meaning of *reason* may or may not be true. It can only be true if the meaning is invariant. This invariance in reason is also demonstrated in the act of observation as it shows agreement between any meaning in the mind and some outside objective comparison. Agreement means no variance. It means something is not relative or cause by an improper form of reason. With invariance in meaning, not only is the meaning correct, but the

3 | The Einstein Illusion

observation fulfills its true predicative nature by confirming the meaning with an outside example. When observation sees something that does not agree with reason, an explanation is in order and you cannot simply change reality objective reality just to make fit the observation. This will only put both reason and observation into objective criteria that is allowed to warp or become variant and relative.

Reason is a subjective process, observation is a predicative process and objective reality must receive the action of observation without being distorted to fit what is seen. When the mind concludes from observation the rational process is backwards. In this you make both reason and objectivity agree with observation. It is the verb that agrees with the subject and not the other way around. The object cannot be distorted to agree with observation. It must agree with reason and only receive the action of the verb which in turn must also agree with the subject. In the other extreme it was Plato who did us a disservice trying to make everything subjective. Thus we have Plato on one side and science compounding the problem trying to make all reality fit into an objective world. You cannot fuse one reality into another. All reality must be incorporated properly in the same manner that language is separated in order to essentially discover agreement and invariance.

The meaning of reason will thus incorrectly fall into two categories. One assumes a certain action and the other assumes a certain conclusion. Both of these assumptions imply relativity. The proper meaning of reason must stand invariant. Properly called subjective reason, it does not relate to separate states or apparent conditions, such as, "that is the reason why" or "this conclusion results from that." In subjective reason, there can be no *this* or *that*. Subjective reason is not a conclusion drawn from a process. Reason is the process and for that matter reason is subjective.

The key in the above analysis of reason lies in the word invariant. For a word to be in a proper reality, the meaning must be invariant, unchangeable, and cannot be relative. The aim in definition, as first taught by Aristotle, is to define words in such a way that they are invariant and unchangeable and free from hypothesis and opinion. Some words do not have clear reality markings, but by studying the most effective synonyms, the meaning can be assigned to the most fitting or invariant part in the language of reality. There are, however, there might be some words that exist in all three realities. It seems that these words are able to transcend the subjective, the predicative, and also the objective. Take the word *truth* for example. Is it subjective? If so, would this not mean that truth comes from the mind with each person having his or her own version? This defines opinion and not truth. Truth can come from the mind as long as it comes from a logical and intelligent process of true reason and not a conclusion drawn from opinion or misunderstood facts derived from observation. In this sense, truth is subjective.

Truth can also be fact or actual existence. This is in the objective sense as the real agrees with the ideal meaning such that a chair is a chair because it has four legs, a seat and a back rest. The object agrees with the definition and thus the objective truth is in the physical chair. Truth has been placed in the objective by science. Observing the chair is mistakenly place in the objective with the chair. In truth, observation is the verb of reality. When *truth* is associated with words like a true observation or a true measurement, truth can also be invariant in the predicate because what makes the observation true or the act of measurement true is in the agreement between a definition in the mind and the objective existence of a chair. A true observation or measurement must show agreement between reason and existence. Here is the essence of a long neglected third reality and the basis of this book. Too often observation does not agree with reason, but we accept the distortion and seek to change objective reality to agree with observation. This is a sin against a sound epistemology. We too often conclude incorrectly by distorting reason to fit what we think the observation means. Science will even draw up equations to fit the observation but modern physics constantly changes both reason and objectivity to fit the equation based upon an assumed understanding about an observation or an observation based upon misinterpreted facts.

Some might suggest propositions to be truth. Possessing a proposition may be a fact, but the proposition itself may or may not be true. Propositions are relative unless they show agreement with reason. Propositions are objective conclusions and need to be tested against observation that shows agreement with reason. In other words, truth can be predicative if it truly shows agreement between the subjective and the objective. This mirrors an axiomatic truth. We now can conclude the following about truth:

> Subjective = Truth is reason
> Predicative = Truth is axiomatic
> Objective = Truth exists

Special relativity has a problem because the observation does not show agreement between reason and existence. It is not that the observation is wrong but reason nd objective reality are distorted to coincide with observation. We need to find existing facts that not only receive the observation demonstrated but also do not distort reason. It should be emphasized that the act of observation is really apparent until agreement can be reached between reason and an objective model. It is the equations that are also apparent and we conclude that these equations depict what we would observe if it were possible. The equations can still be right and even the assumed observation can be correct but our explanations or objective model may be wrong. This is usually the case when reason is not satisfied. Changing the objective model does not solve the problem. Changing reason is only a form of mysticism or psychological motive. Motive can thwart not only religion, but also science. Intelligence is not exempt from this problem. Intelligent minds are often very clever in distorting reason and objectivity to justify what they think we should observe. The many conflicting theories in religion is no different than the many conflicting theories in science.

Religion and mysticism are often destructive when relative meanings are canonized in favor of invariant meanings that are used to warp reality to fit one's personal acceptance formula. Each person must learn to set aside all relative meanings when trying to evaluate reality. A subject, predicate, and object reside in both the human thinking process and in writing. Trying to evaluate reality without an axiomatic predicate is like writing without a verb. The result is an illusion. The absence of a verb in reality is like the absence of faith in reason and existence. In general, philosophy requires the axiomatic creation of basic self-evident statements. This affixes a verb to reality that tests agreement with reason, and objective existence must receive the action of the same. If contradiction results. change the axiom, reevaluate the observation, but never shift the meaning of a word from one part of reality-speech to another. If you do, you simply dump axiomatic reality along with the rational process into a heap of confusion. You cannot turn observation into the truth unless reason substantiates.

Another reflection as to proper meaning occurs with the word faith. From a subjective point of view, faith projects a belief statement void of any reason or proof. This sort of definition means little to the intelligent mind. It appears far too relative. From an objective point of view, faith might be observance or fidelity to one's religion or social norm. This definition refers to some rule, but if the rule be wrong, faith is blind. Again, objective faith must also be relative. Too often we consider faith to be an antonym to reason. Faith more appropriately compliments the rational process. Thus the real strength of faith rests in *the application of self-evident axioms that show agreement between subjective reasoning and the objective model*. The best synonym under this connotation resides in the active word trust. Faith would then be predicative. It cannot be a subjective power as commonly expressed in positive thinking nor can it be an objective list of moral rules. Faith more appropriately balances the equation of reality. Faith or trust in self-evident axioms preserves the rational process by showing agreement between reason and existence. Reject faith as a necessary part of philosophy and *a priori* thoughts are put asunder. Axiomatic faith is the glue that brings the whole spectrum of reality into equilibrium. Faith is that simple obvious knowledge that we come to know out of agreement with the universe. It has no psychological motive; it is patient and shows confidence in equilibrium with existence. It is self-evident and *a priori*. Faith is not blind. It is the intelligent ability to see and then act intuitively with good sense.

Reality pivots on the axis of faith. Faith must keep an eye single to conservation and equilibrium in existence in order to determine if any statement can enlarge our understanding. With true faith as a balance of reality, ideas can be modified and adjusted in order to develop more understanding rather than simply acquire more knowledgeable and more facts. This added understanding could then be used within the process of reasoning as checkpoints to our thinking. The major checkpoint is that all reality verbs, such as observation and measurement, must show agreement between reason and the obtainable facts must agree or receive the action of the same. If the facts are against reason, we must replace or modify our view or consider that our understanding was based upon tradition or even the desire for magic and not equilibrium and symmetry.

As in **Fig. 1-1**, In a total reality equation, objective fact symbolically lies to the right and subjective reason to the left. Faith maintains the axis of the predicate middle as it keeps one's mind focused upon conservation, equilibrium, equity, balance, harmony, symmetry, and has been demonstrated with observation. Just like observation, faith is a true predicate when it shows agreement. Faith is not blind. We cannot expect equilibrium to come easily because traditions and motive inhibit a natural progress. True faith works with conservation and good sense, motive works to establish tradition before truth can surface in symmetry. Faith is the acceptance of sound axioms and motive seeks to exalt itself in strangeness.

One example, using the axiom "justice prevails," shows that true faith can develop an action statement into greater understanding, such that: *Given sufficient time, justice will always prevail through the opposing forces in nature.* It is a lot like gaining wisdom through constant evaluation, study, and contemplation

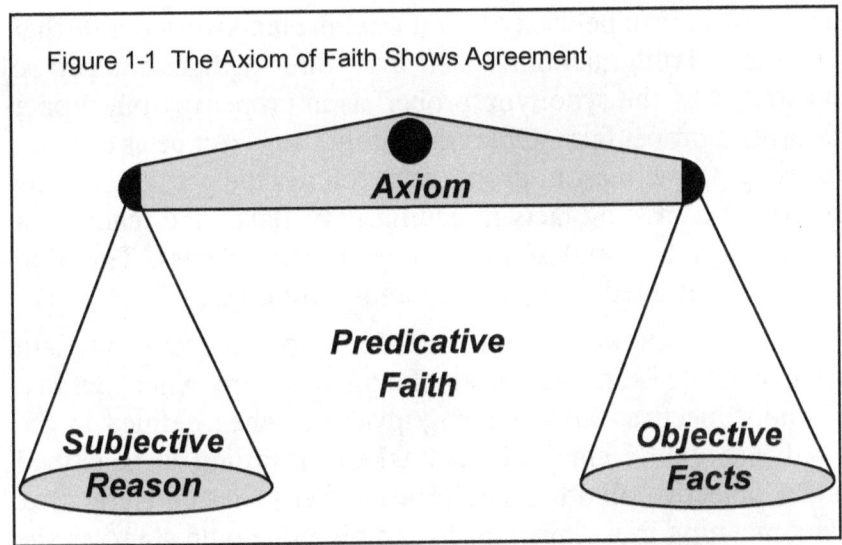

Figure 1-1 The Axiom of Faith Shows Agreement

of many axiomatic statements thus increasing the capacity to reason and eventually increase intelligence. False axioms will simply lead us into intellectual darkness driven by psychological motive and denial. More to the point, false axioms do not hold reality together. The same could be said about false faith. Trusting an incorrect axiom is what we define as blind faith. Faith is like having your eyes closed, but you still process reality in a correct manner, making sure that the axiom in question shows agreement, harmony or supports both reason and existence.

Organizing the mind around self-evident axioms builds a strong statement about the whole of reality. If one argues against the axiom of justice prevailing, that same person lacks faith in justice. The resulting lack of faith will develop a fatalistic philosophy about existence. Reason will then swing without an axiom like a pendulum without an axis. If the axiom "justice prevails" is not acceptable, substitute the word conservation in place of the word justice. No honest physicist will argue against the axiom "conservation prevails," but modern physics attempts to alter this axiom at every turn.

The whole process begins by creating axioms that will show agreement with reason and also indicate that existence will receive the action implied by the axiom. Axioms, once in equilibrium, can be used as a rational presupposition in further processes. Truth maintains itself in all three realities and can be illustrated by the synonym 'proper' as in proper reason, proper faith, and proper facts. Obtaining proper facts can be as difficult as using proper reason. Proper faith follows the path of keeping reason and existing facts in equilibrium. Balancing reality is a constant, active, and sometimes reviewing process. Tradition must be examined constantly by understanding.

As each word extols one definition as more invariant than another, such as reason being invariant when defined in the subjective, faith as being invariant when defined in the predicative, and factual existence as being invariant when defined in the objective; all other definitions must be considered relative. The meaning most invariant for any word should stand as the proper and most effective meaning in science and philosophy. In fact, it cannot be too strongly stated that the origin of each word rightly belongs with the invariant connotation. Tradition, through psychological motive, adds and builds upon all relative definitions, and relative definitions are motivated by denial, fear, and the need for notoriety and control. Relative definitions do not apply faith in good sense or axiomatic trust.

Relative meanings are often canonized in favor of invariant meanings in order to warp one's personal acceptance formula. Each person must learn to set aside all relative meanings when trying to evaluate reality. A disjointed concept about reality as a whole will parallel with incomplete thinking or the inability to think clearly. Even though the mind will often make complex justification using previous assumption, trying to evaluate reality without invariant terms is like writing a statement without the proper parts of speech. The result will be a disjointed statement. Few realize that disjointed thinking is not readily observed without using clear terms in one's own thinking. In general, philosophy requires faith in sound meaning to be placed as invariant at the center of reality before we can even attempt to see disjointed thinking in others. Even though all have the same definition is not sufficient. The definition of each term must manifest the appropriate reality. This is why two Christian religions so often disagree. One places faith in the subjective and the other considers it to be predicative. Likewise, science makes observation objective rather than defining it as predicative.

True axioms (the meaning of terms) begin with equilibrium between reason and existence. Sometimes this is simply intuitive but can be the beginning of a sound theory. The soundness of a theory crumbles, however, when the meaning of words are moved from one reality to another by slightly changing the definition. A balanced reality using axiomatic statements will correct errors and eventually expose a poor theory. Take the study of gravity as a very good example. Some of the most important theories in history began and still occur with this subject. The first, as far as we are to know, was Aristotle's theory. His theory was that the elements earth, water, air, and fire had their natural places in the universe. This statement may appear to have satisfied the predicate reality, but it was based on apparent observation rather than observation showing agreement between reason and existence. It also did not have invariant definitions about the elements, nor did it agree with any prior axioms. For this reason, it was lacking in objectivity on the one hand and clear reasoning about the elements on the other. The words 'natural places' had certain built in assumptions that were not defined invariantly. In other words, the theory did not have a good epistemology because it was a new axiom unto itself without any prior analogues in which to relate. Aristotle's theory was totally empty without so much as a single law of conservation, harmony or balance entering into the theory. The statement 'elements had their natural place' should have been written as 'elements gravitate to a state of equilibrium.' This demonstrates axiomatic meaning without assuming too much. This statement opens the door for a better conclusion. It directs the mind to the proper line of thought without conclusion. It demonstrates how we come to the eventual understanding of natural law.

The Newtonian theory suggested that matter attracts matter proportional to its mass with each equation calculating all mass at the center. This inverse square law states that if you double the distance between two masses, the force of gravity diminishes by four. The statement "matter attracts matter" says a lot about some sort of dynamics, but it says nothing about defining matter itself and does not give us the cause of attraction. It puts matter into the mystical realm as something we should just accept without understanding. Using the law of gravitation we send bodies into space and back, but we do not really understand this mysterious force. Gravitational equations are like blind faith in objective existence without reason and understanding to initiate agreement. We think gravity is intuitive because of observation, but this is an error. Objects do not receive the action of observation in the sense that observation agrees with reason. It is like having a verb and the object but lacking a subjective cause. Newtonian gravity is not a complete reality. Just because we observe gravity to follow a mathematical law does not give us a reason for it. Our faith in the observation and the mathematics is blind without a reason. Do we not attribute the same to religious fanatics. No one has attempted to reason the cause of gravity. We just accept it because of observation and no one has attempted to measure gravity in terms of defining and testing its conceptual workings. Thus we are indoctrinated into traditional concepts without understanding. Just because we see gravity is not enough. We might just think that there are incomprehensible invisible elastic strings that perform the miracle. This is no improvement over attributing God to the miracle. It is blind faith. It is darkness and ignorance. We need a model for gravity that the equations can agree with and existence can emulate.

Attraction at a distance does not agree with traditional definitions of space and time, so we distort their meanings to fit Newtonian observation and Einstein's curvature of light rather than attempt to really explain the workings of gravity and test the theory. We seem only to test the equation that agrees with observation. There is no reason behind the action.

The New Reality | 12

We need a model to explain the reason. Observation justifies the mathematics, but there is no rational model. Our assumed axioms about gravity are not valid until we can suggest a rational model. The only model we have changes the meaning of the terms. By this we have broken the laws of epistemology and good sense. The problem with Newton's law is that it invents a conclusion without any reason. That conclusion says that the action happens at an infinite distance. We are left to invent imaginary fingers, strings or fields reaching out and pulling matter into itself. We neglect the equilibrium needed between reason and existence and we rely on observation to conclude. So what, we say, the mathematics proves it. But we fail to say that the mathematics only agrees with the observation and not reason. The math does not tell us anything about matter or the reason for gravity. We only know that matter, whatever it is, falls inversely to the square of its distance from the earth. We do not know why? Just as Newtonian theory is lacking, modern physics is also lacking agreement with reason. Just to say observation agrees with the mathematics is not enough. It is like something agreeing with itself. Just because mathematics follows logical rules does not prove a rational basis for gravity. We only prove an observational truth, but that is not proving a reason why. Axioms need more and our faith in gravity is far too blind.

Newton did not discover gravity. He set up an equation that said the same thing as the observation. The mathematics is just as apparent as the observation. We obey blindly to the law without understanding. Science mimics the same as the religious believer. The scientist worships the law while the believer worships a god who made the law. Is there really any difference? When faith is based upon tradition or the constant repetition of something, we blind the blind leading the blind and eventually both fall into the ditch. Faith must come from understanding and not blind acceptance. Too many physicists want to be high priests of something miraculous and never allow room for reason. Relativity is the best example.

Until a clear description of gravitational action can be reached, and until a clear description as to the geometry of mass can be drawn, Newton and Einstein's theory stands in the predicate middle ground of observation, action, and reaction void of understanding because the things are only observed while existence is still unreasoned. This does not mean that neither mathematics nor the inverse square law or even the observations are incorrect. It means that nobody knows what gravity really is. Therefore, having a theory about such a thing becomes meaningless. 'Matter attracts matters' or even that 'matter curves' still stands as an empty or blind axiom where both reason and existence are not fully developed.

Modern theory has suggested that space explains the nature of gravity. Mass, both large and small, follows the easiest path through curved space. This relativity axiom certainly mimics that of Aristotle in that matter has natural tendencies, but like Newtonian thinking, curved space runs congruently with the inability to define matter itself. In other words, what would be the geometry of the smallest unit of mass and how can space effect a change in direction? These descriptions remain nebulous and reason is left in the cold. Changing the meaning of space does not solve the problem—it only curve-fits reality to fit observation. It does not explain the action; it only recycles the motion into a new verb. One cries out for a better meaning of the terms and a better explanation of objective existence in order to expand the axiom 'matter curves'. Only reason and conservation in existence can balance the axiom. Term twisting by any other mystical theory slays reason.

The point here is to require a balanced axiom. But when the axiom persists without reason, a need will arise to change the meaning of the terms to fit observation instead of letting reason and existence find equilibrium. Strange theories will lead the mind into strange illusions when reason and a physical model are ignored. It becomes far better that theories magnify and support the three parts of reality by showing agreement between reason and conservation in existence and not depend totally upon observation. Conclusions need to come from the entire process of reality and not only from observation. Observation cannot be a reality unto itself as it has so long maintained.

The New Reality | 14

As with gravity, modern relativity theory has grossly erred by changing the meaning of the terms to fit conjectured high-speed observation and has adopted equations resulting from the same apparent observation. Warped space has become an explanation instead of an objective understanding of things work. What would have been the results of modern gravity thinking if classical axioms were preserved about space, time, and mass, and a new gravity axiom was created that would agree with modern equations yet develop an explanation as to the action taking place without changing the meaning of the terms? This book will build upon the axiom "light curves," but by preserving classical terms, a real description of objective action will come forth.

It should be recognized that observation has governed the modern process. Both reason and conservation in existence were set aside for a predicate reality equation equal to a paradox. The thesis of this book rests on this point: observation mistakenly carried an objective connotation and fostered a change in terms to fit what the eye sees. Holding a consistent meaning in terms, keeping the faith in simple axioms, and develops a clear explanation of objective existence, it is very difficult when one depends so much on the act of observation as objective rather than predicative. To observe existence does not conclude it. Conservation in reason and conservation in existence must be part of the process.

Because science has treated observation as objective rather than predicative, reality without faith in the *a priori* principles has fostered far too many concepts that give false meaning to life. Also, when the meaning of words are turned upside down, it becomes like Humpty Dumpty responding to Alice in Wonderland, "It means just what I choose it to mean - neither more nor less." Existence then becomes a paradox for the lack of faith in the axiomatic. When a strange theory surfaces without a balanced axiom of reason, observation and objective reality, the peaceful mind begins to see that faith without reason or existence without conservation eventually becomes twisted to foster strangeness in order to explain what we see. This process drives strange thinking and strange theory. What really happens? Theory vents one's fear of reality, dumps reason, and rejects conservation in existence. The new reality of predicativism is treated as a single reality and full axiomatic thinking is discarded in favor of a paradox.

A Science for Mysticism

Simple axiomatic meaning seems basic, but why do accomplished minds tinker with a turn of meaning? A new theory poses no real problem as long as the words are consistently defined, the axioms are clear, and experience has substance equal to the definitions. But when theory becomes a paradox because one's subliminal fear trades conservation in existence for a semblance of magic, it can only be said that prodigious thinkers ponder over scientific theory in the same fashion as one ponders over their personal theories about life and death. Faith in pure axioms should be stronger than the cords of death, but too often a suppressed anxiety about fear denies the simple.

Dominating the most eminent minds, the denial of oblivion simply trades faith in balanced axioms for something short of strangeness—all for the sake of a symbolic immortality. In essence, a strange theory can be a psychologically motivated theory. The casting aside of *a priori* sensibility in order to bolster a sense of greatness, simply covers the underside of the theorist's need to become a hero. Only a sound epistemology will prevent such bizarre occurrences by preventing the turning of meaning up side down.

> The first thing we have to do with heroism is to lay bare its underside, show what gives human heroics its specific nature and impetus. Here we introduce directly one of the great rediscoveries of modern thought: that of all things that move man, one of the principal ones is the terror of death.[2]

The terror of death irritates humankind's religious axioms and reminds each of something regretful—our creatureliness.

Both the religious and the scientist cloak their fear of animality and of death with personal belief formulas about God, or a headstrong denial of God. Both lack faith in certain *a priori* axioms about God. The basis of one's theories will rest at the root of one's personal terms and their meanings. When we foster strange theories our belief concepts seem to match well with our definitions of God. This book covers an important point—not to prove God, but to demonstrate that God must be defined in the subjective, the predicative, and objective rather than pushed into one reality or no reality at all. Mysticism must be avoided. Avoiding one of the three realities about God simply covers one's anxiety about life and death. Avoiding reality altogether is just an extreme condition of the denial of responsibility toward meaning.

Essentially, the scientist or layman that denies the underlying fear of oblivion formulates a God resembling the direction of their denial. If one fears the physical, God may become subjective intelligence. If one pushes against reason, God becomes spirit. If one jumps into the pool of objectivity, as a denial of animality, God becomes nature or at worst an idol made of something solid, powerful, and permanent. Many tackle the fear of oblivion by symbolically overcoming God. Isolating Deity into a subjective substance of mind, again as nature itself or a predicate cloud of mysticism can do this. If a man claims to be an atheist, it does not resolve the matter. Every human carries certain beliefs about life and death. Within those beliefs, one will find their God.

Those who accept and understand their personal fear of reality treat God as not only the subjective possessor of intelligence, but the predicate possessor of all good action such as love, compassion, justice, mercy, long suffering, gentleness, kindness and much more; they also treat God as an objective person. This child-like method maintains a full structure and is the best form of sanity and at the same time prevents both the fostering and perpetuation of strange ideas.

The Einstein Illusion

The scientist, no less human than any religious believer, develops a sought-after theory with the same desire for permanence and omnipotence in order to camouflage their fear of life and death. A symbolic heroism walks with each scientist as a soldier of truth ready to exchange the reality of death for the immortality of achievement. This shows most evidently when considering that each scientific authority will have an opposite, but equally unimpeachable view. The religious, on the other hand, condemn the scientific, the physical, and the logical. They simply rely on traditional immortality formulas of mysticism void of sound definitions. The scientist negates or ignores the axiom, "God lives," while the religious zealot cares little about expanding it.

Many modern scientists have affected our religious precepts in ushering in technology. Some had thousands of theories that failed, but it was the tangibleness of these inventions that added a symbolic sense of physical omnipotence. To the determinist, modern technology became an acceptable view toward a physical reality. To the religious, the novelty of an invention could be fondled, and the idol worship given these tangibles was equal to the miracles they performed. Modern technology has produced such an array of peripheral gadgetry that these physical stand-ins for immortality have become the objects of denial for both the religious in common and the scientific superman. As extensions of one's own power and the idols of an unconscious religion, this deterministic view has fused cause and effect into a single physical reality. Faith in religious and self-evident axioms began to fail while inventive science upstaged God as intelligent cause. The classical seer now plays second fiddle to the priest in the laboratory. Science has become the accountant of all miracles and immortality has fallen to the chemist's charge. The entire world is objective. Even reason and faith belong to atoms in motion. There is nothing really subjective or predicative. Reason is not a subjective process nor is observation predicative. All is objective.

The inventiveness of technology constitutes the catalyst in suggesting that all things as physical cause. With all the gadgetry, religion began to adopt physical causality as a tolerate part of reality. But before science had totally expanded this philosophy of objective action and reaction, Albert Einstein (1879-1955) designed a new universe that often quakes both heaven and earth. Einstein's universe played havoc with the newly accepted objectivity, and it ravaged all faith in the classical square of reason. Like a new religion for the modern objectivist, relativity filled the mind with a new mysticism—where reality was neither that of a reasoning mind nor that of unchanging matter, but of a miraculous power ordained upon the believing observer. All of objective reality moved from the objective toward an apparent middle-ground reality of mystic observation.

Beneath the substratum of religion, and even many philosophies, dwells the enigma of mysticism. Mysticism creates a profound awe when strange concepts are declared. These concepts too often imply omnipotence in order to cover the fear of reality. Essentially, the magic of mysticism defies conservation in existence and the theory of relativity does well to mimic this scenario. As a modern mystic, the relativist bewilders the conservation of space, time, and mass through the introduction of observation as an objective reality. Since the inception of relativity, the modern observer-cosmologist has even come to define creation as from nothing—as if things are observed into existence.

This mysticism of a new physics has come to be accepted by proponents of Eastern ideologies and other oriental philosophies, because the observer participates in the cause and not the effect. When Western thought moved away from the causal view to embrace the new observer-defined reality, relativity enhanced the attraction of college students to the field of physics with the same far-out excitement of a magic show. Even the science-fiction writer has adopted relativity by expounding the adventure of time travel. Since its inception, relativity makes it possible for any ideology to keep company with the magic of time. Every protagonist:

> The flat-earthers, the spiritualists, the inveterate believers, all latched on to the apparent enigma of relativity to bolster

their own ideas. Sometimes they rolled many of them into one packet like the author of *Spiritism: The Hidden Secret in Einstein's Theory of Relativity*.[3]

Every wind of doctrine can easily foster a visceral denial of oblivion rather than develop intelligent meaning. In short, the desire of a mystic relativity will easily manifest a defense mechanism against the simple reality of justice in causality and conservation in existence. Although a symbolic form of heroism, mysticism in relativity has developed a cosmic consciousness that slights the simple axioms and gives way to theories that intimidate the power of the common person to understand. Some ignore such concepts because they weigh heavy on the mind, while others fall madly in love with every abstraction. Like mysticism, relativity gives freedom to imagination and a cloak to cover one's apprehension about oblivion. The meaning of causality lingers only as an afterglow while existence persists as the observer sees it.

Mysticism evolved from the original mysteries of the ancients. Designed to create curiosity through subjective symbolism, modern mysticism has shifted from the subjective to the level of observation and distortion. The original intent was for the wise to protect the truth from erosion. Instead, mysticism has become a modern art piece where a personal observer-reality has been created. The original intent of the philosopher-artist becomes meaningless. The message has been lost. Modern mysticism evolved through an attempt to give existence to fear laden precepts. As subjective and objective reality become too bewildering, the modern mystic creates a new reality instead of maintaining a predicate reality to show agreement between reason and existence. In other words, modern mysticism in the new physics has thwarted the great philosophical task of giving consistent meaning to axiomatic words.

> Most importantly, the new physics is offering us a scientific basis for religion.[4]

Mysticism replaces faith in basic axioms for strangeness, exchanges apparent or observed fact for objectivity, and it exchanges reason for a paradox.

> Even more, we must not forget that much of the time, mysticism as popularly practiced is fused with a sense of magical omnipotence: it is actually a manic defense and a denial of creatureliness.[5]

With some people, belief systems block their study of the truth. Though the religious zealot's faith appears blind, it somewhat seems that the relativist denies faith altogether by trying to negate fundamental conservation. Both come from the basic denial of life and death.

> The knowledge of death is reflective and conceptual, and animals are spared it. They live and they disappear with the same thoughtlessness: a few minutes of fear, a few seconds of anguish, and it is over. But to live a whole lifetime with the fate of death haunting one's dreams and even the most sun-filled days - that's something else.[6]
>
> Reality and fear go together naturally.[7]

As the intelligent are subject to the same visceral fear as the common person, mysticism dominates every mind. In science, modern mysticism will breed contradictions and paradoxes to *a priori* meaning, but in religion, mystic ambiguity has been preferred as a sort of mask against facing objective reality. Apparent mysticism in both religion and science simply reject conservation and reason. The scientist rejects faith in *a priori* time and space relationships and the religious use blind faith in order to wash away their fear of reality. The intellectual scientist displays an educated heroism against oblivion by challenging self-evident logic. The religious simply let superstition carry them away.

It was in his third paper, "The Electrodynamics of Moving Bodies," that Einstein set forth his special theory on relativity. In it the classical laws were debased to a substructure revealing the excitement of a more lofty concept; a concept not as fixed in common reason or objective as classical relativity. It was a denial of traditional axioms of space and time, and the common person was left to a universe without clear definition. The new universe seemed neither objective nor subjective. It became a verb unto itself. This new reality was all relative. At any other time in scientific history, Einstein's relativity would have been treated as the doctrine of a false prophet. But many delighted in this new religion of magical cause and effect, where reason without faith in *a priori* axioms promoted a delightful paradox.

Essentially, how any scientist or philosopher gets off track can be found in their epistemology. Why they jump track seems largely psychological. When life and death are inevitable, the spirit holds out for omnipotence and immortality. The struggle for supremacy still dominates the human emotion and the many modern miracles cannot alleviate the fear of eventual oblivion. Man at large still feeds upon pseudo theories of which the ego moves unaware. If the denial of meaning arises as a manifestation of the denial of death, both the fear of subjective and objective reality are denied by the scientist's new reality of observation. The religious simply avoid the subjective and objective by tagging the physical as evil and reason as a lie. The religion in us accepts relativity because it helps to avoid the physical by making all things relative, and it helps to avoid reason by delighting in the magic of a paradox. It does not matter whether the paradox can be resolved. As long as mathematical formulas and terminology back up relativity, no one will ever understand. The common man seems justified, for only the High-Priest mathematician can be gifted with understanding.

Relativity certainly remains as a science for mysticism, something diminutive of a modern religion. As we shall see, the paradoxes can be removed. The equations still stand; but responsibility to conservation and reason becomes the hallmark of science rather than relying only on this new mystic observation.

Effect Without Cause

Turn-of-the-century (1900) thinking considered the human race to be a collection of intelligent animals that like their symbols pure and down-to-earth. Everything must have a cause and each cause must produce predictable results. This simplistic point of view centers mostly upon a highly developed deterministic philosophy, a philosophy of tangibleness where all action resides in the basic physical elements. Even human love was simplified into a compilation of certain atoms systematically arranged to produce the appropriate effect. The laws governing all cause and effect seemed to reside in the elements.

Other philosophies still prosper in what has been called indeterminism. Here, a portion of natural law governing cause and effect reside in the mind as free will. Einstein protested against the artificiality of an indeterminate universe. He preferred a more physical explanation to all things. Little did he realize, relativity was neither deterministic nor was it indeterminate. It was a new reality denying the other two.

As the anxiety about life and death complicates the philosophy of reality, paradox and magic creep into both determinism and indeterminism. This is the problem with relativity. By tossing aside simple axioms, which are easily supported by faith, for a new turn of meaning, accepting events physical events full of paradox seem to be a mere process of creating a psychological acceptance formula. T he relativist is in simple denial of a balanced reality. The determinist's subconscious says, "I am not afraid of this entropic world. If you see, I have rejected all hope in free will just to prove that I am not afraid of oblivion. I accept total annihilation." On the other hand, the subconscious of the indeterminist says, "I need not fear the entropic world, because it is all in the mind. My will can change it. I have the magic within to observe all things and therefore I can build upon my own immortality."

Those who embrace their immortality formulas cannot accept a particular reality placement for the cause and effect world; so, they seek a psychological denial that incorporates a new turn of meaning or the magic of mysticism. Even the modern relativist has sought a personal, observational immortality, where all things reside in the eye of the beholder.

In determinism, both cause and effect are fused into a single physically defined reality. As for indeterminism, both cause and effect are fused into a single subjective reality. And finely, in relativity, both cause and effect are fused into a single observer-defined reality. Human motive simply avoids a full reality of responsibility. Fear dominates the relativist in the same manner as it does the determinist and the indeterminist. Fear becomes an unwillingness to accept a totally balanced reality, a reality incorporating reason, correctly applies observation, and conservation in existence. Instead of overcoming one's fear, each acceptance formula becomes a reality unto itself.

As a new single reality, relativity produces a great many mystical effects. By being born of science, it has power; by virtue of its name, it becomes omnipotent. The believer simply enjoys the entire splendor where nothing seems absolute and the universe lies in the eye of the beholder. The heroic dooms-day prophets tell of total annihilation to bully away their fear; but the coming of a new explosion, a new universe and even a new dimension gives the young believer hope. The classical law is dead, but relativity has opened a "black hole" to a new immortality.

Who has attempted to write a revelation equal to the popular vision on relativity? Who presumes to question such doctrine? Today those who question relativity are finished in physics. According to some, Einstein's theory does not permit tinkering. Its predictions are clear-cut, take it or leave it. It seams, however, that the fiction and nonfiction writers that aspire to relativity still exhibit nothing that can demonstrate any self-evidence. Some writers with occasional comment reveal the obvious, but straightway cloak the oracles of relativity with abstractions. These abstractions are presumed to be understood only by genius, but they simply cover their own ignorance. Other authors call Einstein's relativity a very simple and realistic view, but fail to build a structure beyond apparent observation. The few, who are patient to wait for the justice of God and for conservation in the universe, deserve a far better explanation.

All the rhetoric on relativity must be considered from the addition of the predicate reality rather than the rejection of the subjective or the objective. Relativity is nothing but rejection of reason and conservation in existence. It suggests that the non-real becomes real, and the apparent becomes objective.

Despite the popular notions, space and time still carry certain meanings, regardless of the pedantic games of the theorist. The new physics has often been treated as neither idealistic nor realistic, but a reality based on a single observational reality. Observational physics should never have been a total reality unto itself, because the eye never touches the object. Conclusions about relativity are based on apparent image changes due to time and distance. If any conclusions in Einstein's relativity can be found, those conclusions must show agreement between reason and objectivity and thus develop equilibrium to the whole of reality; otherwise relativity is nothing but a new religion of mysticism.

Why the relativist stumbles on observation as objective, or why the new quantum physicist attempts to push observation into the subjective, bewilders the classical square of reason. Instead of a totally new universe and an altogether new reality, a third reality could just as well be called into relationship as reality's verb; otherwise relativity simply slaughters reason, faith, and conservation. Relativity should compliment both indeterminism and determinism and not be pushed into one or the other.

As long as the priest in the laboratory chants things stranger than fiction, objective reality will remain in the middle age paradox of incomprehensibility. Most importantly, the new observational reality simply becomes an unbalanced reality of observational effect without either indeterminate cause or determinate effect. Like limbo, relativity resides half way between reason and objectivity, waiting for faith in simple axioms to save it.

2. Striking a Balance

> The philosophical consequences of relativity are neither so great nor so startling as is sometimes thought. It throws very little light on time-honored controversies, such as that between realism and idealism. Some people think that it supports Kant's view that space and time are 'subjective' and are 'forms of intuition.' I think such people have been misled by the way in which writers on relativity speak of 'the observer.
>
> -Bertrand Russell [1]

The History of Reality

Each philosophy, religion or scientific theory will carry subliminal design when considering one's justification for a personal immortality formula. As each new theory appears as a guise over a continuous search for a personal elixir of life, we find that nothing ever discovered is absolutely new. The shifting of reality is an age old escape mechanism of the mind. To those who see clearly often are religiously emotional about man's darkness. Those who do not see clearly are more responsive to rhetoric that appears soft, knowledgeable, authoritative, and even humorously justifiable, but the rhetoric is void of logic because of the subtle change of meaning. The student is often of the same psychological makeup as the theorist. At each generation the theory sounds more and more believable and good sense sounds more and more like a worn out song.

Every religion or philosophy has a form of reasonableness, each conspicuously striking the dragon of ambiguity. Each intellect proudly flags its banner for a piece of eminence by waving one reality system over another. Some suppose a reality of subjective idealism; a reality where only an idea can exist. Others tend to pendulum their thinking to a deterministic physical world void of any rational free choice. They accept causality but with no existence outside the physical. And finally the new reality enthusiast lives conspicuously in a life of mystic oneness where logic plays a song of puzzles. The new reality of relativity should not have been set into a single reality of observation, but should have been explained in balance with reason and conservation.

Every philosopher searches for his own elixir. Most philosophies, although they aid in understanding a particular reality, fail to relate to universal thought. Subjectivism, for example, distrusts observed reality by doubting the ability to reduce the out-there-world to understanding. This is certainly a lack of balance. Adhering only to this single reality eclipses a full understanding. They who have a mind to understand let them understand that philosophy has failed by yielding to new word turns that build only upon many single positioned realities and avoiding the whole. The new meanings created for each position will never resolve into harmonic truth. A philosophy must strike a harmonic relationship between three complimentary realities.

From the beginning, the denial of oblivion finds its roots in every ideology. All forms of denial have existed and will continue as a means to mask one's fear of death. Each theory becomes a personal existence formula. Modern thinking was modified by a psychological maneuver when some of the modern physics writers pushed eastern ideologies into a universal subjective consciousness in order to justify a subjective view of relativity. This metaphysics was neither physical nor axiomatic. It represented pure subjectivism and a total denial of faith in *a priori* concepts. Believe it or not, much of oriental philosophy originated from the holistic and not from what is referred to as subjectivism. Modern thinkers have misinterpreted the oriental sage.

Personal reality formulas, as passed down from one generation to another, tend to emphasize one of three realities. Perhaps Plato originated the single sided subjective by neglecting the objective. He perhaps misunderstood Socrates' famous maxim "Know thyself." In this, Socrates felt that knowledge was the highest virtue. Just because knowledge is the highest virtue does not diminish the objective as Plato did. On the other side, Aristotle emphasized the physical with a variation from the Sophists. He maintained that the material world was the real one, and sought to find cause and affect relationships between all things physical. But, should knowing something physical diminish the subjective of knowing thyself or diminish the ability of intelligence to define through observation? The material world is no more important than subjective intelligence and vice versa. Pushing reality to one side or the other is not the solution. Aristotle pushed even the senses into the objective. Some push reality into the middle, as did the Sophists during the time of Plato. They said that all knowledge was relative, that there were no absolute axioms. This made both rational knowledge and objective existence meaningless. Ancient relativity faced the same existential dilemma as today. The modern relativist maintains little difference. Does the modern sophist realize that relativity should show a relationship between the objective and subjective universe, using absolute axioms, rather than evolve into a single reality of paradoxes?

Early Christian philosophers tried to push faith into a form of reason by trading the physical for the metaphysical. Some religious leaders did not accept reason at all as a proper criterion, so the most common result was that the religious philosopher dumped reason altogether and claimed the domain of faith. On the other hand, philosophy has claimed the domain of reason, but tries to make it objective. Neither the religious nor the rational philosopher recognizes that both reason and faith complimented the reality equation as long as conservation in existence is maintained. Faith alone spins a web of magical believing. Without a proper relationship between reason and the objective world, we rely too heavily on 'seeing is believing' and less on 'reason is believing'.

By the 1600's, philosophy was elevated to a position of high authority. With reason the only meaningful reality, thinkers began to stress the natural and physical, which by the 1700's became models for the objective—Newton's book on physics was the most important example. Newton began to put conservation into existence. Only in his theory of gravity was his conclusion based upon observational 'seeing is believing' without a rational explanation. The mathematical formula and the observation it fits did not prove objective gravity. There was no explanation of the action. It did not give us understanding. Newton's gravity was unbalanced. Just because matter appears in an observational way to attract matter does not mean this is the end of reality. Regardless of the observation, the action must show agreement between reason and existence. The real action and reaction that takes place may be on a much lower level than what observation will ever be able to determine. We must start with a better axiom such as 'matter fields interact.' After continued study the axiom could be expanded as 'molecular fields curve in planetary magnetic fields.' Reason accepts that fields interact rather than to conclude that mater attracts matter. The equation has not changed but insight into reality has. This confirms the necessity of keeping reason as the criteria over observed paradox. To say that our minds are limited in understanding paradoxes is simply intimidation as the intellectual bully proselytes strangeness. We shake our heads thinking we are not intelligent. The truth is we have been bullied and intimidated into thinking that the one producing the paradox is more intelligent. Would we say that the preacher of superstition and magic is more intelligent than common sense? If so, the believer is imprisoned by his own fear.

By the 1800's, philosophy turned its attention to various aspects of human experience. Immanuel Kant exemplified a limiting middle ground reality mixing logic and nature. He believed we could only know what we have experienced. Karl Marx emphasized a physical causality mixed with a form of dialectic reason. But his philosophy of dialectical materialism centered more on economics and revolution instead of axiomatic thought. A total denial of faith as an essential axiom of reality produced an empty fulcrum in his thinking. He produced a classless society without a rational God.

It was not until the 1900's that three philosophical approaches emerged. One was the subjective or rational approach based on the development of logic and reason, another was a humanistic approach based on an increasing awareness about the value of man, and finally a physical approach fostered the development of technology and industry. Prior to the 1830, differing philosophies tended to reject in totality the reality not apropos to its specific foundation. After that, both reason and objectivity flourished until the Special Relativity Principle began to merge the two rather than find agreement

Throughout history one philosophy tends to reject or distort a single reality in order to make it conducive to one's basic prejudicial psychological needs. Each philosophy may emphasize some aspect of reality by rejecting either the subjective, predicative or even the objective. The ignorance of a complete reality destroys the foundation of equilibrium and balance. Reason, axiomatic faith, and conservation are complimentary realities. Neither can stand alone. Clinging only to one reality will only mask one's fear of death or rejection. When fear promotes new acceptance formulas by singling out or partially overplaying subjectivism, predicativism, or objectivism without the balance needed, we are in darkness. The following philosophies listed are examples of an positioned or sectionalized part of a total reality:

Limiting Subjective Realities

Subjectivism: Knowledge is limited to self and is not transcendent. Judgments are mental reactions.
Idealism: The real is of the nature of thought.
Indeterminism: Reason is independent of experience.

Limiting Predicative Realities

> *Phenomenalism:* Sensory qualities are the only objects of knowledge and the only form of reality.
>
> *Empiricism:* All knowledge is derived from sense experience.
>
> *Mysticism:* Spiritual intuition is to transcend ordinary understanding.

Limiting Objective Realities

> *Objectivism:* A tendency to deal with things external to the mind rather than thoughts or feelings.
>
> *Determinism:* All events including human choice have material cause.
>
> *Materialism:* Matter constitutes all phenomena including the mind.

Many philosophies overlap more than one reality, but a whole reality must incorporate them all. Reality ought to be a philosophy where objects of sense perception have an existence independent of the mind and independent of one's act of perceiving. Reality ought to define the subjective as a separate reality from both observational aspects of sense and separate from the objects outside the mind. We need three realities and not one or a mystical blending of two.

When one considers the many philosophies since Socrates, one basic motive surfaces in all.

> Existence becomes a problem that needs an ideal answer; but when you no longer accept the collective solution to the problem of existence, then you must fashion your own.[2]

Most individuals, however, are loath to admit their inability to stand-alone.

> This is why almost everyone consents to earn his immortality in the popular ways mapped out by societies everywhere, in the beyonds of others and not their own.[3]

Whether in one's own or the ideas of others, the search for a personal immortality formula deposes one's fear. Deep down, a denial of rejection in life and entropy in death lie buried in unbalanced existence formulas.

The Einstein Illusion

The reader might feel that a subliminal analogy should be far more suited within the covers of a book on psychology or again philosophy. But when one comes to understand the errors in modern physics, that same person will ask, "Why the paradoxes?" Adhering to a single or unbalanced reality formula creates all of the paradoxical errors. In relativity, turning observation into objectivity created a very big illusion. Often related by Einstein himself, relativity was designed to fit observed fact. This placed an epistemological limitation on relativity when observation was considered objective and not apparent. Therefore, the fundamental problem in the philosophy of relativity comes by the shifting of reality. This shifting of the predicative into the objective developed the strange conclusions of relativity. Others pushed relativity into the subjective making it impossible to rationalize the objective.

Since the beginning of the scientific method, observation has been considered to be equal with objective reality. Although appropriate for small localities, observation distorts over large distances and high velocities. This observational distortion of objective reality solidified relativity into a science for mysticism and played havoc with causality. It was not that the equations of relativity were in error, it was that the meaning of the terms used in the equations were turned upside down in order to distort objective reality to fit observation. This was antagonistic to reason.

As Einstein's Special Relativity principle provided a modern denial of causality by warping objectivity, it also subjugated reason to observation through a turn of meaning by Einstein's contemporaries. In opposition to Einstein, some physicists attempted to move the new physics into total indeterminate subjectivism—a religion of observational probability and not a science of conservation. This reality supports the psychology of a subjective consciousness; that of a universe acting upon itself. Anyone delving into these avenues of thought is considered by relativists to be struck by glimpses of many realities. These realities are infinite; because they are isolated into subjectivity. Pushing reality around will never end.

The Age of Distortion

At any point in the ideological spectrum, one will find personal scarecrows. These psychological goblins stimulate the changing of meaning in order to develop strange theories of denial. When distortion of reality runs parallel with the need to establish one's existence or find a personal acceptance formula, the blind instructor leads the student into strange uncertainty.

The lust for magic will not end with relativity nor did it begin with the alchemist's desire to turn lead into gold. Defying conservation in hopes to find a symbolic immortality resided with the alchemist and still lies buried with the relativist's desire to alter space and time. Little did the alchemist know the energy of the stars was required to make gold. Little will the relativist attempt to understand that something unseen in gravitational space juggles the electromagnetic relation? The result: matter curves. The modern scientist's fear of oblivion denies things that cannot be observed, and thus to them space curves by observation.

Order in the universe should be a conceptual reality as well as both observational and objective, but the desire for the universe to exhibit a completely harmonious structure begins to loose ground under the theory of relativity. The new science says, "Put away childish things and face a new reality. Meet entropy, the ultimate chaos, and await the ultimate destruction—a black hole." Like a philosophical bully, the relativist says, "I am not afraid. It's only a bug-eyed monster universe forever folding in upon itself as it expands into oblivion." What common sense begins to see is that the so-called intelligent master of physics actually is trying to deal with his anxiety about death by basking in things stranger than fiction. Many become the intellectual atheist in order to maintain a sense of immortality. Carl Sagan was a recent example.

Throughout scientific history, reason and causality are turned around. Axioms have evaporated by giving faith over to the power of magic, thus killing self-evident epistemology.

Without conservation truth in reason as well as truth in existence will always conjure up a reality of magic. The theory of relativity repeats this pattern by fusing both subjective and objective reality into one observational space-time reality in order to explain the curvature of matter. The results are new definitions such as curved-space and warped-time. If observation and the formulas used in mathematics were treated as predicate realities, the terms would not need this apparent distortion as to both meaning and objectivity. But the psyche of the fearful denies the untouchable, therefore space and time are objective—they warp. Reality should be a philosophy with equilibrium in both reason and existence, but relativity trades it for a physics of observational distortion.

Knowledge is best ascertained if reality reveals itself equally from three points of view. Man's method of constructing a three dimensional world builds upon the logic of first conceptualizing three separate ninety degrees followed by objective three dimensional construction. The apparent observational feedback gives verb action to the concept and the ultimate form of existence to the finished object. Einstein's Special Theory of Relativity causes strain to this reality process. Instead, objective reality now has four dimensions without a fourth for the subjective axiomatic eye. The fourth, neither up, down, left nor right, has a mystical inwardness in the form of a tesseract.[4] Neither the mind conceives nor does the hand create such distortion. Only in the apparent reality of the eye over the process of time's double exposure will a tesseract appear. Reason dictates that a double exposure is really two points in time and not a fourth dimension.

The relativity mind stretchers suggest imagining that the universe represents a flat rubber sheet and if we place a steel ball on the rubber sheet it will distort space because of the gravity of its mass. But the steel ball remains a three-dimensional object whether in a three-dimensional space or on a two-dimensional rubber sheet. This change in meaning of space when moving from the prepositions "in" to "on" does not correlate with any change in the meaning of the mass. Using a three-dimensional analogy to "distort space" draws a four-dimensional concept for which the ordinary intellect struggles to understand. The mind stretchers enchant this voodoo to the simple-minded over and over, melting measurement and objectivity into one reality, but the ordinary intellect returns despondently to a three-square reality of reason, observation, and existence.

Both subjectivity and objectivity are fundamentally based upon three-dimensional space with each plain at mathematical right angels to the other two. To argue otherwise seems stuffily academic. Challenging the truth of the right angle only exposes the fear of being ordinary. As said earlier, the psychological motivation for rebelling against self-evident axioms attempts to establish one's self-existence. Insecurity and fear are at the base of pedantic attitudes. Too much of humankind's objectivity and logic are based upon three coordinates all at right angles to each other.

In order to make any sense of relativity, space and time will have to have more solid meaning. To avoid distortion, time for example cannot be referred to as dimension. The relativist maintains that the equation process in mathematics yields the distortion of space, time, and mass. This seems more a result of epistemological juggling than mathematical prediction. The equation process predicts no such thing as warped space, the slowing of time, or even mass putting on weight. It has been the observational process that yields this distortion and the equation only mimics what has been observed? If there is physical change is must agree with reason. Observation proves nothing. Observation is not understanding.

The language of mathematics as such, represents only a point or function in time. The equation, in a proportional way, describes the action of reality equal to what has been seen. The dependence upon observation to measure out a formula agreeable to this behavior of matter must be required. But it must always be remembered that observation and formulas paint the action of things and do not lay hold of inert existence. Reason predicts existence based upon all previous axioms. Observation only shows agreement at a moment in time. If the equation produces a function over time, observation is not the culprit. It is a lack of understanding the nature of the predicate reality and how things change over time.

Einstein based his theory on sound formulas, those formulas evolved from Maxwell's equation that in turn were developed from Faraday's experiments. Putting it bluntly, Faraday concluded observed action as the dynamic nature of a field or finer substance. Faraday recognized this by simply drawing a corresponding physical model or lines of force to represent the particular nature of a field. It was Faraday's model that will do more to explain relativity than the over dependence upon Maxwell's equations. These equations describe the electromagnetic relation, but do not describe it correctly at great distances or high velocities. Faraday provided an objective model at any distance, yet Maxwell dismissed the objective model. Faraday's objective model is based on reason while Maxwell's equations are based upon observation. It means that Maxwell considered the field to be something like pure energy without substance. Maxwell's view became the seeds of modern relativity. This meant that the dynamics of a long neglected rational model should have guided observation and made sure that the reality equation agrees, with conservation in reason and conservation in existence. Without an objective field that receives the same logic as the subjective model, reality yields a distortion in reason, a distortion in observation, and a distortion in objectivity. How can you write equations about a physical model and yet deny the very existence of such? Maxwell simply pushed objectivity into apparent action. It was as if energy had a force without any mass. This is still the infirmity of modern physics.

The subjective and the objective must have a predicate but you do not need to cast the physical aside when talking of energy. As observation indicates the verb in a thought experiment, equality becomes the verb in mathematics. Thus a mathematical equation denotes a shorthand language for a still photo. Both describe reality at a point in time. A moving picture and a mathematical function describe reality in motion. The photo gives us the essence of objectivity at a specific point. When the photo or observation cannot give us the deeper details, it becomes necessary to create an objective model to test what is observed. Faraday did this, but modern physics does not consider any sort of objectivity to the field. Right or wrong, some objective model showing agreement with reason must be required in order to make sense out of observation and mathematics. We too often think that reason is in every equation. We should understand that the equation simply mimics observation and not necessarily reason. The mathematical process is rational, but not the equation.

Reality's Verb

Observation, measurement, and thought experiments are tools of the scientist. All are part of the apparent predicative reality with relative degrees of precision. More precise than observation, measurement limits dimension to one direction. The observer can move from one right angle to another, but measurement always refers to one and only one mathematical reference point. Mathematics brings observation and measurement together by placing the observer at the zero point of measurement. From this position, space and time appear to change, because what the eye sees in a moving object sees a different rhythm as to what the object sends. The time dependence of light distorts what the observer sees until corrected by transformation equations.

By observing an object such as the sun at noonday, one naturally assumes that the image equals the actual position, but the image transpires through space. The actual position of the sun lies two degrees or eight minutes west. To determine the objective position of the sun, an adjustment must be made that calculates the time it takes light to travel the distance. Too often, the observer treats the image as objective instead of considering that bits of the sun travel the distance. The image is really a dynamic field of photons. Observation then does not equal objective reality. If certain images are observed, the unseen objective factors can only be determined through observation indirectly. The observation must show agreement with both conservation in reason and conservation in existence. That is also the essence of transformation equations.

When using the velocity of light to make a measurement, observation proves relative. Since light cannot be instantaneous nor does it travel a straight line, the relative position and motion of the observer changes the outcome of the observation. If two observers get different images of a moving object, it will affect the equation that describes the object's velocity. Since the equation will only equal what the observer sees when all measurements are made relative to a single observation zero point, adjustments or separate transformations are introduced into the equations to bring mathematics into agreement to other observations. Mathematics then constitutes a symbolical language that describes observed reality from a particular point of view at a particular point in time. This makes mathematics relative to the zero point of measurement or relative to a single observer. This observer is not just any sideline observer. They are standing at the zero point of calculation a particular equation. This point is neglected by all Special Relativity analogs.

Observation, measurement, and mathematics manifest reality in the active or predicate sense. This seems justified from the fact that time enters into observation just as it does measurement and mathematics. Nonetheless, mathematics will balance the whole of reality when the algebraic process shows agreement between subjective reason and objective existence. Just like an axiom, the equation can bring agreement with both subjective and objective reality. Mathematics is nothing more than saying that the object must receive the action of the verb, in this case mathematics, which verb must in turn agree with reason.

Can the object or the subject say they have no need of the other? And can either say they have no need of a verb? Mathematics, measurement, observation, and all of relativity compose reality's verbs. They give past, present, and future tense to what the axiomatic eye sees. But in no way can the verb ever become the object or the subject the verb. Pushing observation and mathematics into the objective truth will never solve the eternal problems; it will only create a bag full of magical paradoxes. If the objective truth of Einstein's relativity principal can be understood without paradox, the unchangeable *a priori* laws in such a compelling new reality must be both logical and exhibit conservation in existence.

The constant velocity of light was Einstein's postulate in determining the Special Relativity Principle. But if light represents pieces of an electron, the first assumption would be that the velocity of light would be at a rate relative to the electron emitting the photon. Einstein, however, insisted that the constant velocity of light travels relative to the observer, because it fits observation. Relativity therefore places every observer at a different point in time in order to keep light's velocity constant. If that cannot be done, simply define space with the ability to contract and expand. The constant velocity again will fit the observation. But this altering of time and space will not satisfy reason or existence.

The only way out of the dilemma comes by defining space as a conceptual unalterable reality, time as constant, and the velocity of light as relative to the magnetic conditions surrounding the observer. Note that I did not say relative to the observer, but to the magnetic conditions. In other words, if light constitutes a piece of objectivity, there must be a change in velocity in order to create the image as seen by the relativist. A change in speed shows agreement with the traditional concept of space. It also shows agreement with observation. And a variable velocity also shows agreement with real objectivity. The only conclusion to consider is that light is not constant in direction or velocity—even in a vacuum. It depends upon the magnetic conditions of surrounding space in which light propagates in.

Modern science, however, assumes that there are no conditions in a laboratory vacuum that would differ from one vacuum to another everywhere in the universe. They are dead wrong. Reason demands that if light is electromagnetic and the magnetic field differs in a vacuum on earth from a vacuum on the surface of the sun, then the velocity of light must vary as it transcends the various magnetic densities in space. The variableness of light would have to be due to the variableness of the magnetic field that light propagates in. We know that matter slows light as it passes through a window pain, but the assumption is that the density of matter or a friction-like event affects the velocity of light. This would not explain why light picks up speed after it leaves the pain of glass. The only explanation would be the magnetic fields are denser within the glass. The angular momentum of light must remain constant, but not the linear momentum. What is meant by angular will be explained later.

The constant velocity of light in a mathematical vacuum carries contradictory meanings due to the fact that light can be refracted by an artificially created magnetic field or even by the magnetic field or authoritatively assumed to be the gravitational field of the sun—the everywhere constancy of the velocity of light in a vacuum was the most foolish assumption physics ever made. General Relativity has proved it to be wrong once you understand that gravity is denser near the surface of the sun. More particular it is the objective magnetic field that is denser and not gravity. A force can be stronger, but objectivity is what yields more density. Objectivity is not stronger or weak. It is dense or sparse. It is energy that is stronger or weak according to the motion of the density considered. By assuming the constancy of the velocity of light, the modern physicist takes objectivity out of light and still pushes observation into objective reality—a total contradiction. The result just seems to be a science full of distortion, paradoxes, and fantasy. If it were possible to discover a greater form of radiation that traveled infinitely faster, relativity would prove apparent as it should be, and all the popular notions about the new physics would vanish as an illusion in the night.

Conclusions are often based upon assumptions due to the meaning of a given term. If the meaning is not clear or nebulous in nature the theorist usually adapts personal meanings into the thought of a new process or even changes the meaning of a term to fit a curved representation. When mathematics incorporates these distorted terms into its equations we often make conclusions about reality that the quantitative calculations do not dictate. Mathematics only deals with the numerics and not the terms. Change the meaning of a term and essentially you change the philosophical outcome of the equation without changing the syntax, operators, or quantities. The term *space* is a term that is a general expression that defines length. Its meaning is often changed to something that warps negating the theorem that the shortest distance between two points is a straight line. Now the shortest time between two points is a curve or some relative change in direction. This concept is mistakenly applied to space.

Samuel Dael

3. The Square of Reason

> There is a deep-seated faith (no other word will suffice) dating back to Greek times that the universe exhibits order and is basically simple. Whenever any facet seems to grow tangled and complex, scientists can't help searching for some underlying order that may be eluding them.
>
> Isaac Asimov[1]

Absolute Basics

Agreement between concepts within the mind and real existence outside the mind should yield an underlying order in the universe. When scientists search for truth in the physical world, they do so by the act of observation and measurement, but what they see must show agreement between the conceptual terms in the mind with terms attached to the objects being observed outside the mind. Since observation and measurement often change over time, the scientist attempts several observations and measurements in hopes to eventually reach agreement or at least an understanding cause of any particular conflict. Accepting a paradox between the ideal and the real is not rational, intelligent or meaningful. It is obscurity.

Faith in this agreement drives scientific thinking and can be focused more clearly by defining the relationship of equality, distinction and proportion in the rational process. For example, if two objects are not equal, there must be a definition attached to each object that will explain the difference or the distinction. The absence of these definitions in meaning is really the absence of reason. This is because reason defines things free from contradiction. To accept a paradox rather than explain any distinction is darkness.

Figure 3-1 illustrates the use of distinction between the meaning of the word *small* and the word *large* when associated with the objective 1 meter and 2 meter rods. Distinction yields inequality. Observation demonstrates this inequality by showing agreement between the meaning of the conceptual terms and the objective labels determined by measurement.

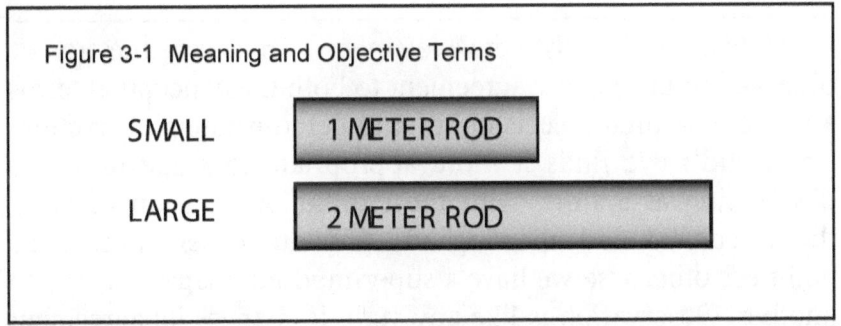

Figure 3-1 Meaning and Objective Terms

If we were to change the conceptual terms to *good* and *bad* from *small* and *large*, observation will not show agreement because the new conceptual terms do not relate to measured objectivity. What we are doing is distorting meaning. Modern physics does not attempt to distort meaning in this way, but it does distort objective reality by moving observation into objectivity in order to explain what is observed. This will become apparent as this argument progresses. Now if the two rods were equal, as determined by measurement, there would be no need of conceptual or *spacial* terms. The absence of a distinction requires the absence of a conceptual distinction. Two small rods or two large rods do not have meaning because there is no distinction. Two equal rods would only reveal confusion if conceptual terms were used. Equality, in other words, is the self-evident absence of spacial or conceptual terms.

By the introduction of a third distinction in **Figure 3-2**, the apparent verb agreement, in this case (*is*) changes between the spatial concepts *small* verses *large*. What was large in Figure 3-1 becomes medium in Figure 3-2. In other words, observation

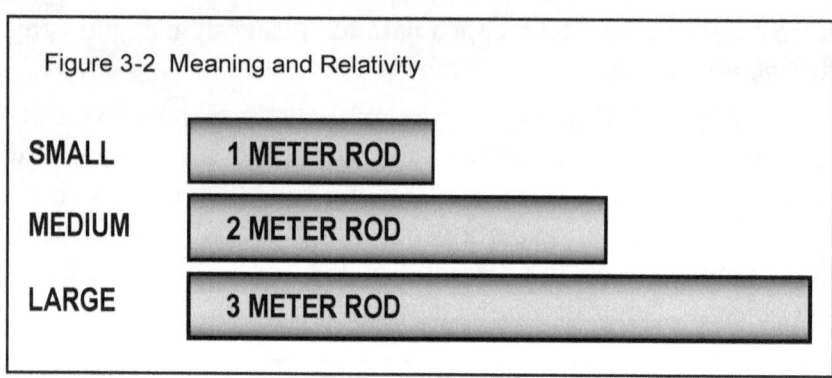

Figure 3-2 Meaning and Relativity

is relative to the distinctive elements. In short, the act of observation must show agreement to both the conceptual terms and the measured objects in order to demonstrate a distinction. The mind's eye finds it more appropriate to make only one observation at a time, with the current observation carrying the agreement and allowing any previous observation to be nullified, otherwise we have a superimposed image contrary to another. Observation is like any verb. It changes its agreement with each observation as a verb changes it's with each sentence. Observation is an act of doing something. It is relative and not conclusive from one moment in time to another.

The meaning of the word *large* does not change. Also, the objective two-meter rod does not change without a physical cause. Though it was designated the "large rod" in the first distinction and a "medium rod" in the second, objectivity will not change with a new observation. If the spacial meaning and the real objects do not change, what does? Simply, the act of observation imparts apparent but not actual change. Spacial concepts never change except by redefinition and material objects do not change without objective cause. In no way can the act of observation affect either the concept or the object. Observation is a verb of reality and it cannot change either the subjective concept or the object itself. Observation is relative to each case as a verb is relative to each sentence.

Mathematics is also reality's verb because it is relative to what is being measured, but measurement never changes the object. The modern theorist argues that observation and measurement are participators in objective change. This has been exemplified in the act of measuring a hot solution with a cold thermometer. The temperature reading of the hot solution would not be equal to the temperature before placing the thermometer into the solution. This would be true because the thermometer cools the hot solution, ever so slightly. But if the thermometer were the same temperature as the solution, the thermometer would have no effect on the solution except for possible friction. For this reason, the effect of the objective thermometer depends upon the distinctions of the term *temperature* and *thermometer* as to a solution. Temperature is related to measurement and a thermometer has nothing to do with the act, rather it is another object. In this sense temperature is a reaction and thermometer is a cause. We make the mistake of thinking that measurement of temperature is the cause and it is not. It is the physical thermometer that affects the physical solution.

Distinctions are the objective elements, including the molecular thermometer, and have nothing to do with the act of putting the thermometer into the solution. The act of measurement is predicative; all the interacting elements are objective. The modern thinker fails to keep these two realities separate. In order to measure the effect the thermometer has upon the solution separate from the act, this would require many different measurements to show the distinction between the temperature of the thermometer and the solution. In other words, several measurements and the appropriate transformation equations make it possible to eventually reach a measurement of the solution where the thermometer has no effect on it.

The difference between the spatial or conceptual definitions of the units, the act of measurement and the objective measuring device altogether incorporate three different realities. So when talking about space, you need to decide which is the reference—the term, the act of observation or the object. The object is relative to change over time and the act of observation may also be relative because every observation is slightly different, but the conceptual terms cannot change because each has a precise definition. We mistakenly say space changes because the ruler changes. Understanding the rigidity of invariance of spatial definitions reveals that twentieth century relativity errors when it treats space as objective—a thing that warps. It is like saying that the definition of an inch warps or changes in meaning. This is incorrect. The spacial terms or units of measurement are definitions and are not objects. Definitions do not change—only the objects over time have cause and effect change. Just because the measuring device changes over time does not change the definition of the conceptual terms. Change pertains to the magnitude or value of the units and not to the definition as one would say an inch shrinks at high speed. The ruler might shrink, but not the definition of an inch. If we look at the ruler to determine an inch, we must assuredly conclude that observation is not any more fixed than the ruler over time.

The essence of a definition is that it becomes the subjective concept derived by a previous axiom or a complete thought that clarifies a distinction. Distinction means to define. The subjective then is invariant once the meaning is accepted. Observation becomes the act or predicate of seeing and is invariant when the object equals the concept. When the object varies, we must then introduce a causal reason why and use a transformation equation to explain the cause. We cannot change meaning simply to explain a change in objectivity. We cannot accept objective change without understanding why. To do so is not intelligently rational. Equality must eventually arise out of observation or an explanation is in order. Changing conceptual meaning is not an explanation.

The essence of equality is derived by dividing the concept by the object, including any transformation adjustment. When we get *one*, we have equality. Equality demonstrates the invariance of the object from the concept. If the distinction between the concept and the object is zero, equality arises in the meaning of the term *one*. Equality shows agreement. Quantitatively, if the value is one, the essence of this "oneness" means *equality*. An infinite number of distinctions such as big-small, left-right, black-white etc. gives us many variations in terms to choose from. Equality signifies no distinction, at one with, constant, symmetry and absoluteness. Equality stands as the only invariant, present tense, axiomatic reality used in mathematics.

In **Figure 3-3**, equality also expresses in geometrical form the symmetry of a right angle.

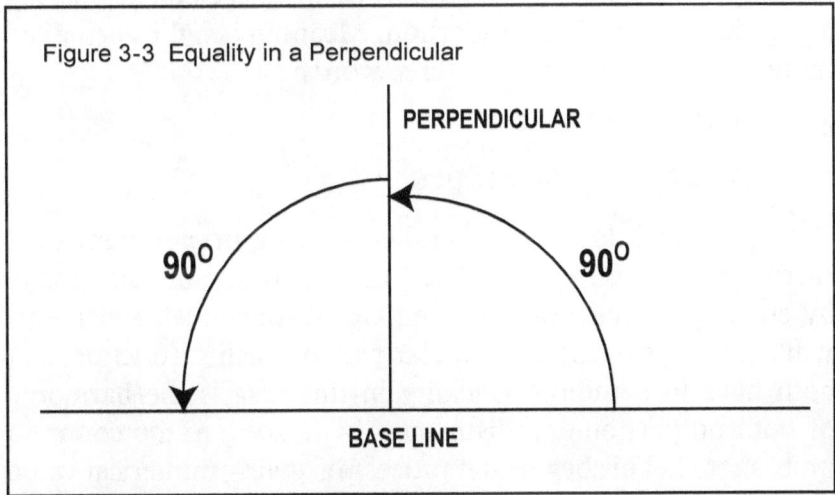

Figure 3-3 Equality in a Perpendicular

Drop a perpendicular to a base line and the left angle equals the right. Any angle less than or greater than ninety degrees depicts a distinction. Without a single equality of angles out of an infinite number of distinctive degrees, geometry could not exist nor could the right angle create a meaningful concept for space.

The variables in mathematics in conjunction with the operators yield a precise shorthand language that says the same as what can be said with words. Equality carries the same meaning as the verb "is" in language. Because relative verbs denote infinity, referring to the past and future tense, it becomes necessary to find one absolute or present tense verb to develop a logical system of thinking about the objective world. The one non-variant and only absolute found in the predicate reality is equality.

Equality In Proportion

Measurement between two distinct items, using a common unit, can reduce a distinction to unity. In other words, a five-meter rod and a ten-meter rod have something in common. They can both share the same spacial unit. In the above case, that unit is "meter." As distinction relates to the total difference, equality relates to a common unit. Two distinct elements sharing a common unit yield a proportion. Meaning can thus be nailed numerically to the following three words:

> ONE defines EQUALITY
> TWO defines DISTINCTION
> THREE defines PROPORTION

With a little Pythagorean-like thinking, the next numerical word "proportion" generalizes the meaning of distinction by creating a total ratio between two distinctions or elements while equality forms a particular part of each distinction that both have in common. Equality in this case is the harmonic of both proportion and distinction by looking to the common units shared. This begins the process of giving numerical value to terms used in physics. Each higher numerical word will also incorporate all preceding axiomatic words as well as incorporate a number of concepts equal to the number assigned. A very rigid method indeed, but meaning would unlikely change when words are thus nailed with numbers. Interestingly, not only does proportion incorporate a distinction of two elements and the equality of one common unit, there are three kinds of proportions. They are numerical, directional and definitive.

A simple example of a numerical proportion compares a bowl of fruit with a bushel. If a bowl of fruit has four apples and eight oranges and a bushel of fruit has thirty apples and sixty oranges, the proportion between the bushel and the bowl introduces the numerical ratio of "two to one." A numerical proportion does not have direction, but it does have magnitude. Understand that magnitude pertains only to the numerics and not to the subjective terms "apple" and "orange."

A directional proportion illustrates the equality of angles with opposite directions. Unlike magnitude, which has infinite ratios, such as 2 to 3, 1 to 5 etc., the distinction in a directional proportion resides in opposition. Opposition means distinction in a spacial sense. **Figure 3-4** depicts this relationship between two, one hundred eighty degree angles—equal in angle, but opposite in direction.

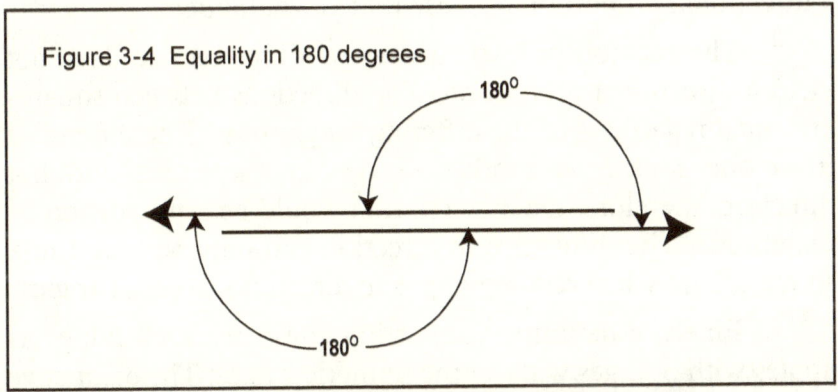

Figure 3-4 Equality in 180 degrees

Between the equality of the angles set at opposite directions, a square and rectangle are also proportional. Even in **Figure 3-5**, the associate angles still show in a parallelogram.

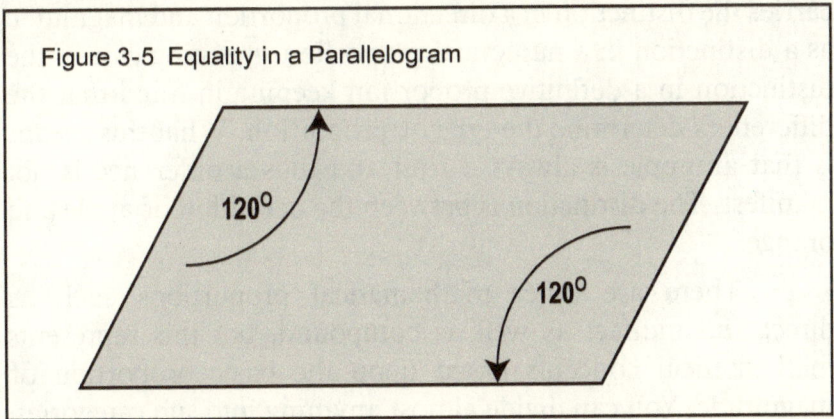

Figure 3-5 Equality in a Parallelogram

The geometrical figures created by opposite angles are proportional in direction and not magnitude. This does not mean that the figures do not have magnitude. It means that the proportion or relation carries the nomenclature of direction. The numeric size of a square is meaningless to the conceptual term *square*. Essentially a proportion carries the nomenclature of the differing aspects and not the elements in common. In general, if the distinction differs in magnitude, the proportion yields magnitude. If the distinction differs in direction, the proportion yields direction. Direction differs in the above-mentioned angles, but they share the same magnitude; therefore, it is called a directional proportion because that is what differs.

The relation of two squares of differing size does not yield a directional proportion. The directions between squares are equal regardless of the differing magnitudes. The difference from one square to another resides in magnitude and not direction; therefore, the new relation would be a proportion of magnitude. The difference in direction can only be found in a single square when relating opposite directions to equal angles.

Finely, a definitive proportion compares such things as apples with oranges without the numeric values. The distinctive characteristics between an apple and an orange will vary, but a definition in common classifies the two as fruit. The third term "fruit" remains constant or equal in definition whether one speaks of apples, oranges, plumbs or grapes. Just as direction carries the distinction in a directional proportion and magnitude as a distinction in a numerical proportion, meaning carries the distinction in a definitive proportion keeping in mind that the differences determine the type of proportion. What this means is that an apple is always a fruit and thus a difference is not manifest. The distinction is between the definition of apple and orange.

There are other mathematical proportions such as direct and indirect as well as compound, but this represents mathematical concepts based upon the basic proportion of magnitude. You can divide almost anything into sub categories. The above argument is broader and more conceptual. We can now look at the proportional relationship of space and time? What do they have in common and how are they distinct? Are they distinct in magnitude, direction or meaning? A space/time proportion has two names. The basic one is speed which yields a proportion of magnitude. Now space and time differ as to reality, but this is the definitive aspects of the proportion—something that will be illustrated. As to mathematics the difference is in magnitude. Both space and time share the same directional line segment eliminating speed from being called a proportion of direction. The difference between space and time is definitive as to reality and distinctive as to magnitude. It is these two areas that are of concern and not the concept of time differing in dimension. Time is not a fourth dimension because it shares the same traditional x coordinate with space. Time is a derivative from space and not a dimension.

Space and time differ only in magnitude and definition. For example, if a ten-foot line segment were marked off in equal one-foot units and if the tick-of-a-clock units are marked off on the same line segment, the time marks could align with every other foot mark if the speed was two feet per second. Speed, then, does not indicate a directional proportion because no distinction in direction can be found. Both share the same line segment. The difference comes in the magnitude of the numeric values compared—such as 2 feet to 1 second.

Just as the magnitude of speed equals space/time, two speeds can be related in the same fashion as two squares. This is exemplified in another space/time term called *velocity*. In the most rudimentary form, velocity has the same space/time analysis as speed. But velocity, by definition, becomes a directional proportion because two speeds differ as to direction as well as magnitude. Even if one speed equals zero and the remaining speed has any given angle, the result is velocity. Just as two directionally proportioned squares result in a proportion of magnitude, the result of two speeds of magnitude determine a velocity of direction. Textbook terms refer to this as a vector. A vector has both magnitude and direction, but the present consideration attempts to separate the numeric values from the directional aspects in order to analyze them separately. Once understood, it would be all right to generalize this new space/

time term called velocity as having direction because the two speeds share the same calculated vector of magnitude. Velocity in its rudimentary space/time form must simply be speed. It takes two speeds in different directions or one speed relative to a specific zero point to produce the concept of a shared vector velocity

Whether one speaks of velocity or speed, as in 2 feet per 1 second this proportion of magnitude cannot be found in the conceptual or definitive terms *feet* or *seconds*. It would be the numeric values assigned to these terms that carry the distinction in magnitude, for when the terms *feet* and *seconds* are removed from the relation, the numerical ratio of 2/1 still stands. Mathematically, space and time are numerically proportional, but not directionally proportional unless we use vector addition of two velocities relative to each other.

This means that the terms *space* and *time* are distinct as to meaning, distinct as to magnitude, but not distinct as to direction. Modern relativity thinking fails to illustrate this point. Just because space and time share the same line segment does not give time its own direction separate from space. The basic reason for this is that real time has no direction, is not subjective or objective. It is predicatively based upon the equal frequency or action of matter. It is also predicatively based upon a directional component of space. Time is derived. It is a derivative and not a dimension.

As explained in speed, both space and time share the same numeric values over one direction. Time shares the same single direction in order to create a direct proportion of magnitude. Space and time are fundamentally equal complimentary parts to the whole mathematical statement of reality. If you cannot see the nature of this reality, with mass the only objective component you will certainly miss the mark and fall prey to paradox. Just because a clock might move along the same line segment as the object of study does not equate time with direction nor does it equate time as objective. Clocks are objective—not time. Also knowing the conceptual difference between magnitude and direction should clarify this as much as knowing the difference between time and the time piece. Magnitude does not define direction and direction does not define magnitude. Direction is subjective and magnitude is predicative. You cannot relate them as having something in common as *inch* is to a *foot* or *yard*. Time is not a subunit of space like inch is a subunit of both foot and yard.

For this reason, it would not be the numeric values of relativity equations that distort reality; it is changing the reality of the terms by pushing time into the subjective as in direction and space into the objective as something that curves. A correct meaning of the terms must show symmetry and balance. Separating the distinction of magnitude from direction can do this. In the numerical sense, the first meter in a space/time proportion must equal the second if both were to be placed in the same direction from a zero point. The same applies to time. One movement of a timepiece must equal another if both the timing device and conditions are equal. If it cannot be done, there would be little point in depending upon the algebraic process. What would it mean if the fourth inch was shorter than the seventh or that the eightieth-second was slower than the third? It would mean mathematical chaos and conservation would be destroyed. This illustrates that the invariance lies in the reality of the terms and not with the values or magnitude of those terms. The definition of meter never changes, but the meter on an objective rule will change over time. It is so important to differentiate between the subjective or definitive concepts from the objects used to measure physical reality. Some would argue that you cannot separate time from the timepiece. This voodoo thinking has caused the inability for modern thinking to separate action from mass—thinking it to be convertible rather than different parts of reality. I think the problem is a psychological one due to the lack of conceptual intelligence and an enormity of linear intelligence—a byproduct of modern education and the decadence of the classical square of reason.

> Using the inch for a unit of measure to signify ten inches, or decimal equivalent, becomes very significant with

the pyramid and other structures. But the Egyptian inch differed from the British inch. Postulated by John Herschel, an early British astronomer, a unit half a human hair's breadth longer than the British inch could be based on the polar diameter of the Earth. An early British Survey had fixed this diameter at 7898.78 miles (by taking the mean of all the available meridians measured). This translated into 5000,5000,000 British inches or an even 5000,000,000 inches if the British inch were slightly longer.[2]

Defining the ancient inch in such a way fixes the meaning. Also, understanding that there was a decimal equivalent in constructing the Great Pyramid illustrates that over time the definition of the inch could have been corrupted. A modern definition of the meter uses a similar definition such that the meter was intended to equal one ten-millionth of the length of the meridian through Paris from pole to the equator. However, the first prototype was short by 0.2 millimeters because researchers miscalculated the flattening of the earth due to its rotation. Still the definition became the standard. This illustrates that the subjective reality or definition of an inch differs from the objective reality of a small measuring device created to very closely equal the definition based upon a very large calculation. When relativity then considers the warping of space, we have to consider the warping of something very large in planetary terms. When the ruler warps by observation or by objective cause, we know something is amiss, but the warping of a solar system—not so easy.

When the relativist takes conservation out of both space and time and gives time dimension while pushing space into objectivity—something that warps is an infatuation with paradox more than reason and conservation. It is a cultural-religious problem dealing with responsibility rather than intelligent understanding. Again, the terms space and time define a numerical proportion as to measurement and share the same spacial segment. It is like two realities sharing numbers marked off on a single line. The relationship is the same as the relationship between the subject and the verb. When the verb agrees with the subject there is no distinction. Measurement shows agreement with space as any verb should show with the subject. Mass on the other hand receives the action of time as it should.

Just because space and time will share the same numeric values of magnitude of a particular single direction as in velocity does not mean space and time both have direction? Think of space as having magnitude and direction, but time is only a quantity of magnitude because it only shares the magnitude of space and not its direction. Just because an object moves in a specific direction does not equate time as moving in the same. We could also draw the same proportion of space and time so time would be symbolically placed at 90 degrees to the distance traveled as in **Figure 3-6**. A diagonal can now represent velocity but the actual velocity of the object of study does not have the same direction as the diagonal of the chart. It appears that we have a directional proportion as if we are comparing two aspects of space as a directional proportion. We are really comparing the two magnitudes of both space and time and the direction of space has nothing to do with the diagram. When we think of *inch²* we do not consider direction for when we say square inches we refer to the magnitude of the number of inches and not the direction they are measured. We only know that the 90 degrees is essential in order to symbolically represent in

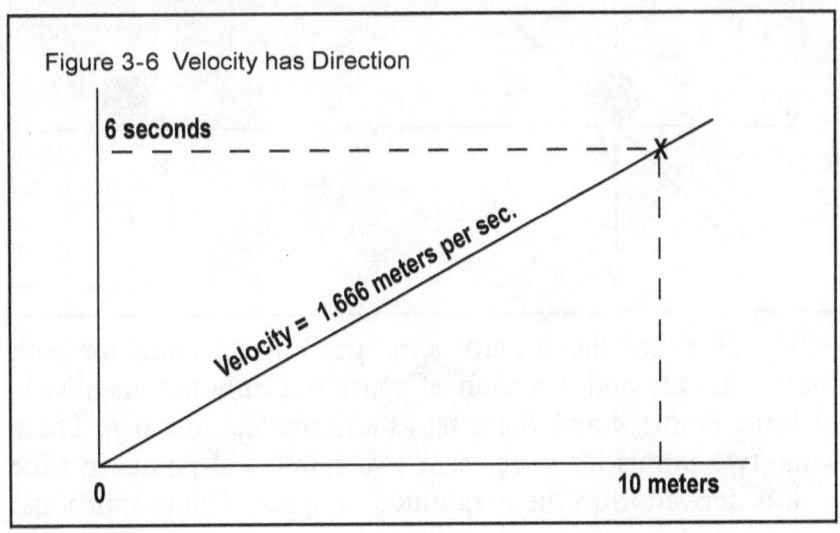

geometry what actually is magnitude only. The use of *time²* is also magnitude only.

A space/time relation can get very messy, depending upon how many directions and times are involved. Perhaps this justifies the confusion. But the terms must be kept clear, even if the numeric values and dimensions are tossed about. Space and time are distinct. Breaking things apart has particularized this point. Relativity jumped track by mixing the numeric values with the definitions and completely neglecting conservation. Since conservation pertains to the terms and not the numeric values, both space and time are constant in the equation and must be constant as to the whole of reality. Relativity can only be applied to the magnitude of a space/time proportion and not to the reality of each term. Relativity accepted a change in terms and thwarted the logical process, thus advancing well over one hundred years of incorrect philosophical thought.

In **Figure 3-7** velocity or acceleration are conceived as an arbitrary point traveling with an object in study. The time measurement and the space measurement run congruently by symbolically sharing the same numeric values of the *x*-axis.

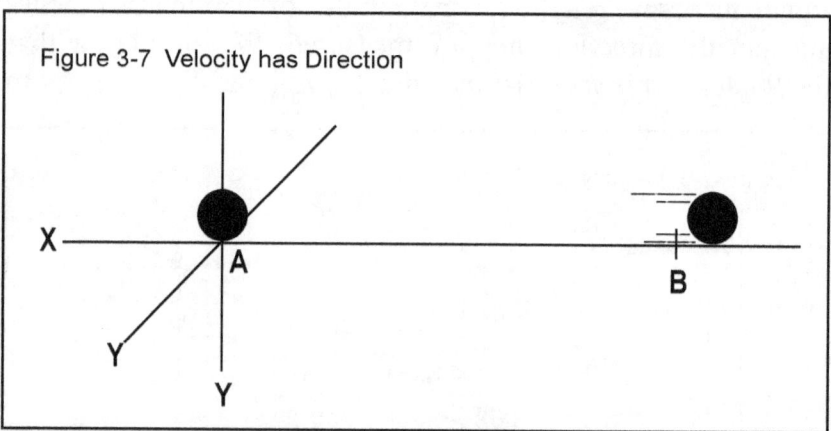

Figure 3-7 Velocity has Direction

This axis has a zero point and an end point for both the magnitude and direction of space but only the magnitude of time. Points *A* and *B* are set as a numerical distance. These same two points also represent two numerical points in time as it is derived from the magnitude of space. The proportional relationship of the magnitude of length and the magnitude of time yield the meaning of velocity. Time squared, or the second derivative of time is called acceleration.

VELOCITY = Length / Time
ACCELERATION = Length / Time squared

Do we think that time here has two dimensions? No! It is a change upon a change or a change in velocity. Direction has nothing to do with acceleration unless it is a change in direction. But this is like two velocities with one vector representing a net single direction. A vector is a special addition of all forces and velocities that combines every direction into one and only one. With space you do not add dimensions—you multiply them. You can have an infinite number of directions all adding up to one net vector, but you cannot have an infinite number of dimensions. Geometry will not allow it. Even when you cube time it is adding up the various forces into one vector. Just as there is a difference between direction and magnitude there is also a difference between direction and dimension. When the physicist uses time in terms of a dimension it is only the figurative expression for a single vector of space/time. It thwarts the student from learning the principles of vector analysis.

Space and time are not the only elements of the equation process. A proper definition of mass must also be understood. Although mass must be used as a distinctive component in a proportion, mass itself is a proportion. According to textbook nomenclature, it can be expressed as follows:

Mass = Force / Acceleration

Mass cannot be particularized as easily as space or time. It comes from a complex relation. In fact, it takes mass to define mass. A simple example can be demonstrated through the following equation:

Mass = (Mass x L / T square) / (L / T square)

Mass appears on both sides of the equation. For this reason, mass seems difficult to define in the particular. In general, mass requires a constant force over an accelerated distance as in.

Force = Mass x L / T square
or
Force = Mass x Acceleration

Force is a vector, which means that it has both direction and magnitude. The essence of force is in *mass* but although mass has magnitude it does not have direction. A three-dimensional ball might imply direction, but this is not what the mathematician measures. The quantity of mass is calculated as if the entire quantity is at the center. The size is only a spacial dimension that distributes the magnitude of mass to a larger volume. Mass (thought of as a certain number of atomic particles) resists motion in any direction. If the resistance to force over a given directional distance remains constant, it follows that the number of particles must remain constant. If any particle was cast aside and the force remained constant upon the remaining body, acceleration would result. For this reason, mass remains constant as long as the acceleration is constant.

MASS is proportion of magnitude
FORCE is a proportion of magnitude and direction

The direction refers to the term *force* and the numeric values of a specific number of atomic units pertain to both mass and force as the magnitude is shared. As illustrated earlier, it is the difference that determines whether the net result of a term is magnitude or direction. In the case above direction is the difference and thus force has direction. Let us also analyze *work*. The textbooks refer to it as a scalar quantity or a proportion of magnitude. Here is the equation for work:

WORK = Mass x Length square / Time square
or
WORK = Mass x Velocity square
or
WORK = Force x Length

Both force and length share direction and thus the difference is magnitude, thus work will be a scalar quantity of magnitude. Again, the determination of a vector or directional quantity and a scalar or quantity of magnitude is determined by the remaining reality after all shared realities in a proportion are cancelled.

The Square of Reason | 58

Note, however, that in the first reference to *work* above indicates that both space and time are squared. This gives an interesting clarification to popular relativity. If space has three dimensions, then time should have three derivatives. Perhaps this is where the relativist mistakenly gets the concept of many dimensions. Unless the math can show it there is no reason to assume anything. The third derivative of time, however, denotes *"control"* and when used with mass, it becomes *"mass control"*:

CONTROL = Length x Time cubed
MASS CONTROL = Mass x Length x Time cubed

> In any case, because this control factor is not in textbooks, we should give it special attention... Clearly, a change of acceleration is what we mean by the word control,[3]

Called the "free agent" or the unpredictability of the system, control becomes the third derivative of velocity. This argument is given by Author M. Young, a helicopter engineer.

Time does not have three dimensions nor does space in the case of *velocity square* in *work* imply two dimensions. To square space in work yields the second derivative of the first order of magnitude and not a second dimension. The square of a piece of tile on the floor and the square of length in *work* carry different abstractions; only the numeric values or magnitude remains the center of concern. The two-line segments in work share the same axis; therefore, *length²* becomes scalar only. This provides us with another reason to define *work* as a proportion of magnitude only and not direction. Time squared is also magnitude only and has nothing to do with direction.

The ideas above are more conceptual than what one learns in mathematics today. In fact it has been the tendency for modern education to concentrate on the obtaining of rule assumptions about data more than understanding through conceptual analysis—better put as understanding. We think that concepts are for advanced minds having some sort of upper graduate factor. This assumption is actually incorrect. Concepts pertain to dimensional factors rather than the regurgitation of information in terms of magnitude. Children can be taught this naturally, but the system wants to judge in terms of a scalar score rather than inspire a conceptual ability. Some have conceptual intelligence more than others. The problem comes with the fact that conceptual minds are slower while linguistic data minds are fast. In this respect, the slower minds are weeded out in a fast-track educational process. The result is that many linear students become linear educators that propagate the same and all future learning promotes minds of magnitude rather than conceptual or directional minds of relation. Relativity is the very by-product of linear minds claiming a paradox as a solution. It tends to intimidate others into thinking they are more intelligent, but actually it reveals conceptual inadequacy. Einstein was conceptual because he had to have others do the mathematics. His Special Relativity Principle was none other than the result of making observation symmetrical with empty space and the constant velocity of light. These two assumptions are incorrect. There is no such thing as empty space and logic says there must be some sort of variable tension of light to make us thing that space is curved and time is dilated. Relativists that followed Einstein did not correct the error of the constant velocity of light and still assumed the wave properties in empty space even though Einstein concluded particle characteristics. To this day the problem of empty space and the wave and particle duality has not been resolved. The problem comes from excessive data mining rather than conceptual analysis.

Essentially, it takes a conceptual mind to maintain conservation in space/time and also in mass. Linear minds just accept what was told them. What seems so bizarre is that all the twisting of terms in behalf of relativity is done in the name of mathematical conservation. In other words, the equation as a whole must be conserved in order to fit observation. The space/time components are allowed to become relative in order to make the observation work. Since observation cannot see what is causing the distortion, the relativist twists the terms with the quantities to conserve the equation. Does this resolve the problem? Before this thesis ends, the answer will be no!

The Law of Conservation

A law or axiomatic meaning endures when it can stand the test of time. The law of conservation speaks this harmonious truth very well. Subjective reason demands it; predicate faith in equality necessitates it; and most important, there would be no existence without it. Conservation logically reduces all reality to unity by bringing everything to a level of equilibrium. Equality is the essence of logic and mathematics and makes all understanding possible. Conservation is an *a priori*, common sense concept that has intuitive insight yet it is human nature to deny it for a view of paradoxical magic. Within objective reality, conservation is manifest by the value of four, with conservation of mass and energy each divided into distinctive aspects. Here I introduce the essence of the number 4.

4 defines CONSERVATION

I. Conservation of inertial MASS
Conservation of gravitational MASS

II. Conservation of momentum and inertial ENERGY
Conservation of angular momentum and gravitational ENERGY

I hinted the two types of mass earlier, but more particularly two types of energy are also considered in conservation. General relativity considers inertial and gravitational mass to be different manifestations of the same thing, but the mind can abstract the two separately and define different types of actions—one being resistance to change (conceptually called inertial) and the other being gravitation (conceptually referred to as tension or curvature). Inertia is manifest in the scalar number of atomic units while curvature is manifest as a directional aspect due to acceleration and force—the greater the curvature the greater the mass. The equality of the two is that both have their own quantities for calculation and thus both have magnitude, but gravitational mass has a directional component in its derivation from force and acceleration. The conceptual mind looks at what is going on as to direction while a scalar quantity is sufficient for calculation. But keep in mind that gravitational mass does have a directional force just as much as a hammer against the many atoms in a nail. With inertia the action and reaction are opposite, but with gravitation the action and reaction behave more like the magnetic and electric field at right angles. The difference is how we measure and also how we conceive. There is no difference in objectivity. The four concepts of conservation are spacial and measurement in much the same way as a thing has direction and magnitude.

Objectively speaking, both inertial and gravitational mass appear as the result of the same substance, but one is measured in reaction to a linear force while the other is measured in a reaction to an angular tension. The apparent action of each separates the two. No physicist questions the difference between linear momentum and angular momentum. Although the mass might be the same, the momentum is calculated with an additional right angle vector in angular momentum. If there is a difference between linear and angular momentum, there is conceptually a difference between inertial and gravitational mass. While momentum demands the existence of inertial mass, angular momentum respects the existence of gravitational mass. The difference then is how the action is defined.

Does this mean that mass exists in two objective forms? In scalar terms this is not the case, but in conceptual or directional terms there is a difference. We can understand that inertia is relative to the quantitative mass and the directional force applied. In gravity the force changes according to the inverse square law. In inertia the force is action with an equal and opposite reaction. What is lost in the action is conserved in the reaction. *In gravity we have acceleration due to a constant angular torque of matter's electric fields upon a diminishing magnetic field until such a point there is no torque in outer space.* You will not find this in any text book because it is conceptual and the reason for which this book is written. In summary we have two forms of mass and of necessity we must have two forms of energy. Once this is understood we can better understand why both mass and energy are conserved and are not convertible. Contrary to relativity, mass and energy are not converted from one to the other. Rather angular energy and linear energy are convertible just as momentum and angular momentum are convertible. Gravitational mass, better put as angular mass, is converted to inertial or linear mass and back again. The same applies to energy as liner or kinetic and angular energy as that bound in matter. Before this thesis is finished it will be proven that the classic equation $E=mc^2$ does not suggest conversion. Both the mathematics and the concepts are misapplied. Ignorance abounds because of it. Energy is like a verb to reality and mass is like an object. They do not change places as purported by relativity physics. That is the essence of this thesis. Both are conserved independently.

In passing all of particle physics and onto field theory, the universe has thus been demonstrated as a composite of at least two fields—the magnetic and the electric. Combined into a composite unit, the two fields fold in upon themselves. The result yields inertial mass such as one would measure in a particle. When an external field to the particle affects this inertia the result will be measured as gravitational mass. Before field concepts can be illustrated in the creation of matter, we need to understand the nature of conservation—the primary cause of matter's discovery.

The use of conservation in physics started with the French chemist Antoine Laurent Lavoisier (1743-1794), who was commonly called "the father of modern chemistry." Lavoisier maintained that in every chemical reaction in a closed system the total mass remained unchanged. This law has since been called the conservation of mass. Later during the study of nuclear reactions, when more precise means of measurements were obtained, the law of conservation was still considered immutable when even a disappearance of mass was detected. A stubborn but logical attitude to save the law manifested itself in the universal concept of conservation of energy. If any mass disappeared from a closed system it was calculated as a form of energy, but should be expressed as a form of light. Both light and mass carry energy because both are forms of mass.

Energy is mass in motion. You cannot say how much energy is in a stationary particle unless you are attempting to describe the angular momentum of the constituent parts—thus the concept of angular energy is more appropriate than rest energy or mass.

Because of the apparent disappearance of mass, relativity physics fused conservation into one mass/energy reality to save the law of conservation. There was no need for this if energy could simply be mass in relative motion. The apparent weight of a falling mass may be nothing more than the microcosmic angular momentum of the molecular body's electric field working at right angles to the angular momentum of the greater body's magnetic field. This angular momentum of both fields, when converted to the linear momentum of the falling mass, reduces the angular momentum of the fields and increases the linear momentum of the falling mass. Energy means energy, whether it resides in a small angular moving field, or in the large linear falling mass.

If a stationary mass lost some of the microcosmic angular moving particles (light if you will), the "apparent" loss of mass, formerly measured as gravitational, could now be measured separately as inertial mass moving linearly from the larger gravitational mass. What was part of the whole angular system, now measures as a separate inertial mass having a certain velocity and having colliding power at impact.

Regardless of the definition of the mass/energy relationship, conservation stands as the only absolute worth saving in the new physics. Some say, however, that within the dark reaches of a black hole, all the laws of conservation collapse. It behooves some to believe in a white hole on the other side to save the law of conservation. Such pedantic logic leads us to question the value of observation. Without some absolute constant, such as conservation, the equilibrium in reason, faith and existence says nothing. We must have an eye single to and continue our faith in conservation of both mass and energy.

The scientist views conservation with the same conclusion that the audience questions the magician's trick.

The inquisitor will always ask, "How did you do that?" And the magician may answer, "Oh! The hand is quicker than the eye." If conservation was not so fundamental we would never ask, "How come?" Conservation works in logic, measurement and physical causality. Accept every trick with reservation, because logically there must be an explanation or cause and effect relationship. The magician, however, depends upon the mystical thoughts of the audience: "How wonderful it would be to obtain control over the elements that I may make something from nothing." Such a wish challenges the law of conservation and could be considered a denial of our own limitations and a wishful glimpse at immortality. Man fears the inevitable, men fear the unknown. This is where we marvel at the magician. The spirit of intelligence speaks of logic, yet the spirit of fear says, "Let the magic scare away my fear of life and of death and give me a false sense of immortality by saying that all things are in the eye of the beholder".

Relativity's basic assumption defies the logic of "how come," but it has a bag full of tricks to captivate any audience. Warped space, time dilation, distorted measuring rods and strange inertial forces demonstrate a few of the quicker than the eye illusions. Some will marvel at this science fiction trickery, but others still cry out, "How do you do that?" Regardless of the magician's genius there must be a trick.

Guard against mere illusion, unless you can admit to simple fantasy and entertainment. Enter back stage and see what really happens as the relativist waves his magic wand. Hold on to the only absolutes that serve modern science: equality, distinction, proportion and the law of conservation. The law of conservation expresses equality in physics and also within a complete thought. Like distinction and proportion; conservation reduces all measured reality to unity and full understanding. Intelligence can then lean upon reality with a true sense of the square of reason.

4. The Geometry of Mathematics

> Mathematics is a shorthand language in which each symbol has a precise and agreed-upon meaning. Once the language is learned, we find that it is only a form of English after all.
>
> -Isaac Asimov[1]

The Subject of Space

The axioms of equality, distinction, proportion and conservation exhibit concepts with precise meaning. These *a priori* self-evident concepts require a deep-seated faith in the cogitative ability to define all things properly. Modern writers often call some of these laws generalizations for fear that they may change with further research. This overly academic attitude powers a false perception about reality and particularly about observation. It should be remembered that the self-evident laws evolved just as much from reason as from experience. Reason interprets conservation, experience justifies it, and faith maintains it.

The modern interpretation of Einstein's relativity principle defies reason by saying that space, time, and mass are not conserved individually. Only by converting one reality into another does the mass-length-time ratio remain constant. Sound thinking objects to this, for when observation changes, faith asks, "What caused the change?" If you know the change explicitly, then all things are conserved individually and not collectively. But if you do not see the cause of the change, but only the total effect, this empty observation says nothing. You are apt to

The Geometry of Mathematics | 66

think that space/time can be converted to mass and vice versa. Without faith in fundamental axioms of equality, distinction, proportion, and conservation, reason becomes a paradox, the act of seeing fosters a new reality, and verbs become the objects.

In mathematics and physics, words should not be used metaphorically as the word "page" in "a page of history." As some words precisely define, others are used in realities they literally do not manifest. Space is a perfect example. Physics gives precise meaning to words such as velocity, but since space does not have clear reality demarcations, the new physics simply tossed it about and redefined space as something objective—a thing that warps.

As attempted earlier, conceptual distance has been defined differently than that of numerical or scalar distance. Conceptual space, or the mind's eye, stands as an independent reality from the numeric values of distance. This does not mean that space originates in the mind. It only means that the mind apprehends or conceives a rigid frame without the aid of numbers. The numeric values, or measurement on the other hand, give magnitude to the concept. Though the senses act over a period of time and so too does measurement, conceptual space does not. Space remains an instantaneous graphic to the mind and must therefore be conserved. In the mathematical equation, the logic of the mind's eye operates in the individual variables of space. The measurement assigned each variable assumes correctness even though the measuring device will alter in time; otherwise mathematics and logic are fallacious. Existence can change over time and thus indicate the need of another observation where new measurements must be made. This conclusion does not defy logic or the nature of mathematics. It only means there is an explanation yet to be discovered when time is a part of the equation. The mathematical equation says, "The measurement varies but this is not due to a lack of conservation in space, time or mass individually, but it is due to the fact that observation and measurement are relative from one point in time to another. Time is relative when the time piece changes, space is relative when the measuring device changes and mass is relative because it is a dynamic substance of fields. It is the objective aspects that change and not the subjectively defined units. Mass is conserved if you can count all of the photons and their individual size that enter and leave over time. Space is conserved if you can maintain the definition of an inch, meter etc. Time is conserved if the objective clock does not change or is not affected by outside sources such as magnetism or gravity. Conservation is maintained in space, time and mass or we have to explain the reason why. You cannot move one reality into the other in order to explain observation such as space being objective rather than subjective just so you can say, "It warps." It is more appropriate to say, "The measuring device warps."

The common man carries a precise concept of empty space until the scientist attempts to make space a thing that warps. Space cannot be the mass any more than the subject can be the object. The only way to describe the objectivity of space is in the word "nothingness," but warped nothingness boggles the mind. The physicist's concept of space should more logically exemplify a metaphoric example—something on the order of light refraction rather than curved space.

Intelligent order satisfies the law of conservation and maintains the absolute nature of space. Metaphorically speaking, if you steal a word from its subjective orientation, literally a void remains without a word to name it. Intelligence is anti-entropic. It conserves what the eye cannot see. If, for example, you reject the word "space" as a subjective concept and give it tangible characteristics, the mind will still conceive the infinite void that has no name.

Why is it so important for the physicist to make space an objective reality, something that warps? Perhaps space also moves, stands still, slows down and decomposes as other objects do. Leave conceptual space to its own, for intelligence requires something absolute from which to compare measured reality. The concept of space serves this purpose. Without it the mind desperately searches for some spatial equality to compare its definition with the object through observation and measurement.

It must be remembered, the scientist measures objective reality and not space. Subjective space provides the absolute concept tailored for the terms in the mathematical equation.

> As a matter of fact, Euclidean and non-Euclidean geometry appear to conflict only if we believe in an objective physical space which obeys a single set of laws and which both theories attempt to describe. If we give up this belief (in a physical space), then Euclidean and non-Euclidean geometry no longer rival as candidates for a solution of the same problem, but just two different mathematical theories.[2]

We must give up this so-called physical space. It does not work with reason. The need for attention getting carries every writer away. And in drawing new definitions, they often exaggerate or change the common ground meaning of a word to turn out a piece of work; namely, to oppose in a competitive way a more commonly defined version. With such denial, fear appears as an escape from the realities of life and a failure to accept the spirit of truth. "But its all semantics," retorts the deluded, "for all things are relative." Uselessness always appears when the mind finds comfort in imagination or when a mistaken observation appears as existence. To think objective reality changes at will by selecting a different observation point falsifies reality and treats observation as objective rather than relative to a position as it should be. Observation and measurement are relative and not space.

Space cannot be the material and space will never be the physical event. If the material object appears as a two-dimensional sheet of paper, intelligence refers to a two dimensional space. If the object bends as a flat rubber sheet, intelligence can no longer refer to two-dimensions. It must picture three.

As mentioned earlier, squaring a coordinate does not necessarily imply two directions. The length in "work" for example, does not describe the area of a two-dimensional square. **Figure 4-1** will illustrate:

A to B defines the path of motion. The work done equals the line integral of the force over this same path. The second and third dimensions, defined by y and z, do not enter into the

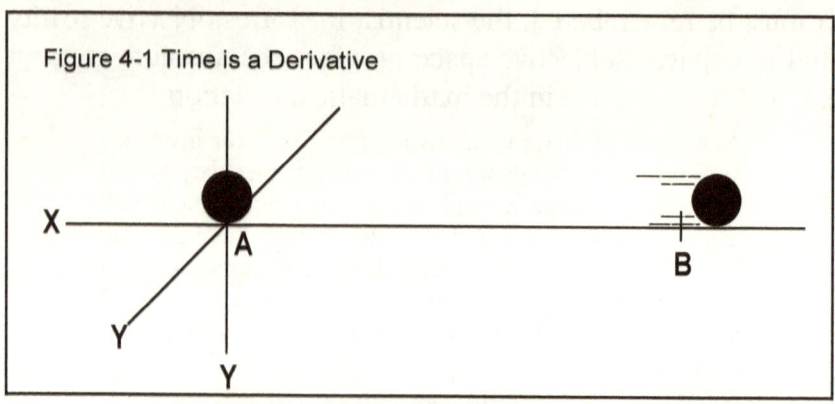

Figure 4-1 Time is a Derivative

calculation. The first "length" defines the force applied over the distance A to B and L^2, or the second length defines the work done over the same x coordinate. L^2 then defines a derivative of magnitude and does not define a second dimension. There is not a single case in mathematics or physics of an equation that gives L a greater value than three dimensions. All values greater represent congruent "lengths" along the same single coordinates. Space must be limited to three-dimensions and the distinction between subjective dimension and mathematical derivation should be made clear.

It has become fashionable to talk of four dimensions in Einstein's relativity with time becoming the forth. More recently, however, eleven dimensions have been postulated. This attention getting reveals itself as nothing more than several mathematical magnitudes, all of which will share a similar line segment. With this understanding, time cannot warp space because space will never be the object. Only mass will be affected in time. Time gives meaning to mass in the same way the object receives the action from the verb. Direction comes from the subjectivity of the mind while time represents a form of magnitude and tells nothing of direction.

Time Has No Direction

If an object moves 10 meters in 20 seconds, time brakes up the 10 meter length into 20 equal parts. Whether measured in ticks, vibrations, a swinging pendulum or any object in motion, time divides the one-dimensional distance into equal units. The equality of each spatial unit can be checked separately regardless of the timepiece. The inconsistence of a timepiece does not change the fact that time in the equation persists as a constant. The clocks we use to measure time carry physical properties, but nothing about the essence of time can be measured as physical. Time arises from the motion of things. Time is measurement dependant. Time is apparent. One might ask, "How fast does time move?" Apparent reality cannot be measured in this sense. Objects move but time has no speed because time embodies nothing without the object. It is like energy that also embodies nothing without the object. It is only something the mind can sense but not objectively see. The sense concept of time is constant, but the observation of time is nothing but the observation of a time piece ticking away. One might conclude that time is subjective. Like space perhaps in some sense time is, but the difference is in the action of objectively and not the direction. Here again we are talking of something scalar rather than directional. The only thing directional about time is the direction that the object travels—this is according to space. Space applies a subjective reference while time applies a predicate or active scalar reference to the object.

No equation defines time alone. Can one refer to an object as being 20 seconds, or an inch as being one minute? Can a verb stand without a subject or an object? Realities should be treated separately as in the case of dimension and derivative, concept and measurement, inertia and curvature and even measured time from the timepiece.

In the equation of speed, the object has no dimension. The length defines the distance traveled, but the same coordinate also defines the same length during a time interval. As algebraically shown earlier, time is directly proportional to space in magnitude, not direction. More precisely, time supplies a congruent magnitude, not a fourth dimension. Velocity only has direction in geometrical terms as one object traveling at a particular angle to a plane of origin.

The idea that time seems relative comes merely from the premise that measured space is relative at high velocities. It is not space that is measured. It is the moving object. So if the object changes, the ruler also changes. Is the ruler space or is it objective mass? Also, the Special Theory of relativity concludes that the necessity of time dilation comes from the need to balance the equation using warped space. If space is not warped, time will be constant. Mass relativity resolves from the same assumption. Vary the space and the need to vary time, and mass levels the equation. Why do we not admit that things change—including the observation and the measurement while space, time and inertial mass (as particular units) are constant?

Understanding the immateriality of both space and time requires the conservation of mass. The timepiece may run slow because it will change with a change in magnetic or gravitational field; therefore, a clock is not constant, because the clock itself constitutes objective reality. But just because time measures reality through an objective timepiece, it does not make it objective any more than measured space such as the terms inch or meter are objective. It describes the action of the object and not the object itself. Time shows agreement between subjective reality and the objective universe and, for this reason, it is derived rather than converted or changed.

As time derives its meaning from the motion of things, the measurement of time comes from breaking up space into equal units. The measurement of mass also derives its magnitude in a similar fashion. The derivation of mass comes from the measurement of both space and time.

MASS = ACCELERATION x Length / Height

In the extended form we have the following:

Gravitational MASS = (Length / Time square) x (Length / Height)

Figure 4-2 will demonstrate that the above three length measurements and two time measurements, all lie along the same vertical axis. All can be represented by one direction of space.

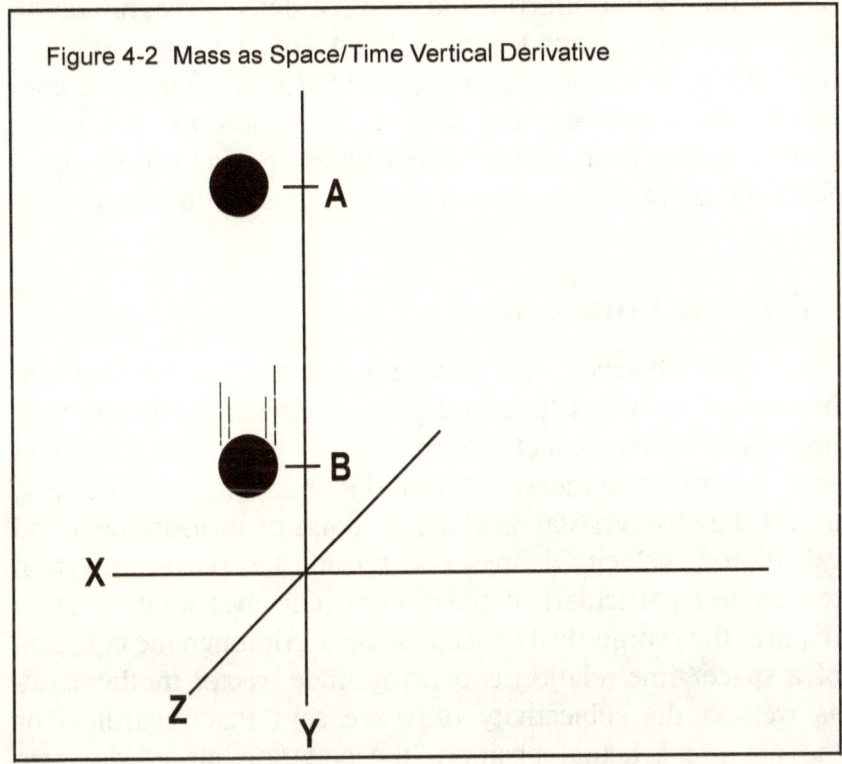

Figure 4-2 Mass as Space/Time Vertical Derivative

As before, the x and z coordinates play no part in the calculation of gravitational mass. Even mathematically, mass has no apparent direction because the entire quantity exists, in calculated form, and at one point called the center of mass. Although every mass has dimensional size, the equation can only calculate its magnitude as a point along one and only one direction.

When considering that time and mass can appear as derivatives of one single direction, the proportional relationship exists in magnitude only and not as a dimension. When intelligence uses mathematics to understand indirectly the existence and action of objective reality, it does so by extending a coordinate into one dimension and finding a numeric relationship between the extended magnitude of space and the magnitude of time. Each of the three spatial magnitudes can exist by themselves, but time in the equation cannot. It needs one of the spatial directions to create a derived magnitude. In this sense time is not a fourth dimension, but one, two and three derivatives of a single magnitude of space. Time becomes neither the object nor the subject. Time becomes the active pattern maker over a single line segment having no direction. Such should be the conclusion that time has no direction.

All Things From Zero

The observer often takes for granted that the constant motion of an object follows a path relative to the position of observation. The geometry of mathematics says that a constant velocity cannot be measured from the observer's point of view, unless the observer stands at a zero point of measurement and calculation. Velocity defines a vector and has precise direction relative to a particularly defined point. This measured direction requires the geometrical concept of zero. Although the equation of a space/time relation is of magnitude, vector mathematics as well as the subjectivity of space says that regardless of the observer's actual position, the equation places the zero point behind the object's direction of travel and thus a sideline observation would actually differ from the equation. We mistakenly think that the observer determines the velocity. The actual mathematics calculates the velocity from a measured point of arbitrary beginning and not the standing point of a sideline observer.

Relative to a sideline observer, the velocity of a vehicle changes constantly. Negative during approach, zero for an instant and positive during the recession, the vehicle also accelerates relative to a sideline observer. The mind, however, compensates the irregularities of motion by placing an imaginary rigid reference frame along the path of travel and assumes that the zero point of origin connects rigidly to the reference frame. The mathematical equation does the same. In every experiment the observer happens to be a bystander and cannot be included in the measurement. If the observer says that an object has a constant velocity relative to his or her standing point, that point must lie between the moving object and its straight line origin—the zero point of all calculations.

The equations of mathematics have no knowledge of the observer as a bystander. The equation will assume the arbitrary point of zero to be the point of observation. Remember that mathematics limits the calculations to only one reference point. The geometry of mathematics demands the rigidity of this point. And when the imaginary observer does not reside at the zero point of measurement, a thought experiment may very well differ from the calculation in the equation.

An object will move in more than one direction at the same time, but each direction has its own zero reference point. For instance, a billiard table has a mathematical reference point for each moving billiard; indeed, each billiard moves in only one direction relative to its own point of origin on the table. But if the table rides upon a ship, each billiard moves in two directions at the same time. One direction is relative to a point of origin on the table and another direction is relative to a point on the earth. Consider a third wave motion of the water as a third direction. And to complicate matters, the rotation of the earth, its orbit around the sun, even the motion of the sun, all this adds to the total number of relative motions of each individual billiard ball. To complicate the situation, a zero point as a physical mark may move during the time interval it would take to measure the velocity. Does that change space?

Consider three observers, all at different locations relative to the motion of an object. To one observer the object could be moving towards, to another observer the object could be moving away, and to a third the object could be passing. Each observer would describe the motion differently, even though the motion remains constant regardless of each observer. Unless skilled mental and mathematical adjustments can be made, observation remains completely relative.

Finding a point of reference that will satisfy all observers equally creates an obvious solution to the problem. Mathematically that position stays at an arbitrary zero point along the direction of travel. Velocity requires that the magnitude of time as well as length be measured along the object's path of motion, for both observation and mathematics reveal the same results from zero. The limitless number of observation points stands against the backdrop of only one mathematical zero point. As in a twelve-inch ruler, the twelve inches lie in a direct line from zero and not from a sideline point.

The many thought experiments used in explaining relativity fail to consider the difference between zero point measurement and observation. In the Special Theory of relativity, the thought-experiment and the sideline-observer erroneously endures as an equal to the mathematics.

The concept of using light as a method of measuring distance and thus velocity must place the observer at a zero point of measurement and not at the sidelines. The use of radar demonstrates a good example. Imagine a moving object such as a truck along the highway. By general observation, you stand at a distance many feet into a field. The truck seems to move at a constant velocity relative to you as the observer. You sense, however, that as the truck approaches a position at right angles to the road, its speed moves more rapidly. And as the vehicle recedes, the speed becomes less apparent.

If you want to be sophisticated, perhaps the use of a radar gun would give you a more accurate picture. As you point it in the direction of an oncoming vehicle you notice that this device has in the view-finder a number that registers the velocity of the moving object. "That's terrific," you say. As you endeavor to hold the equipment steady and on the mark, you notice that the digital number reads -55. You interpret that as fifty-five miles per hour relative to you. The radar sends a signal to the moving truck and when the signal returns a small computer within the unit calculates each successive signal to determine a change in time of travel. If the time diminishes at a certain rate, the conversion yields a negative number. If the time increases with each successive signal the rate equals a positive number.

A strange thing then begins to happen. As you move the radar unit more rapidly to keep the truck in the viewfinder, the digital number begins to drop. As it approaches zero, the angle of motion of the radar unit will shift very fast. You almost loose sight of the moving object in the viewfinder. As you gain control, the digital readout quickly climbs to a positive number until it finally levels off at +55. If the truck travels at a constant speed of 55 mph, why does the readout progressively change from a negative number to a positive number? And to confuse things more, why does the velocity register zero at a moment when the radar unit pans the fastest? Both observation and measurement make things appear relative because the observer is at zero point for radar transmission and not at zero point to the equation of velocity.

In **Figure 4-3**, the zero point for the radar unit defines the point of observation. The radar unit does not have the capacity to determine the angle of motion. The internal circuitry only interprets that the object moves toward the radar gun with a diminishing velocity followed by a motion away from the observer with a progressive velocity. Radar works on the principle of the rate of change in distance. So also does light.

Radar from the observation point interprets point *B* as one impulse later than point *A* and not as a linear distance from zero. From *A-B*, both the distance traveled and the radar distances between pulses are equal, but from *B-C* they differ. At first, the distance between vehicle and radar observer decreases more rapidly during the first ten seconds. During the second ten seconds the distance appears shorter. The rate of change in pulses and not the distance itself determines the velocity from the radar's point of view. If the position of the radar unit could send signals from the origin of the moving truck's zero point, the rate of change in distance would be a constant +55 mph.

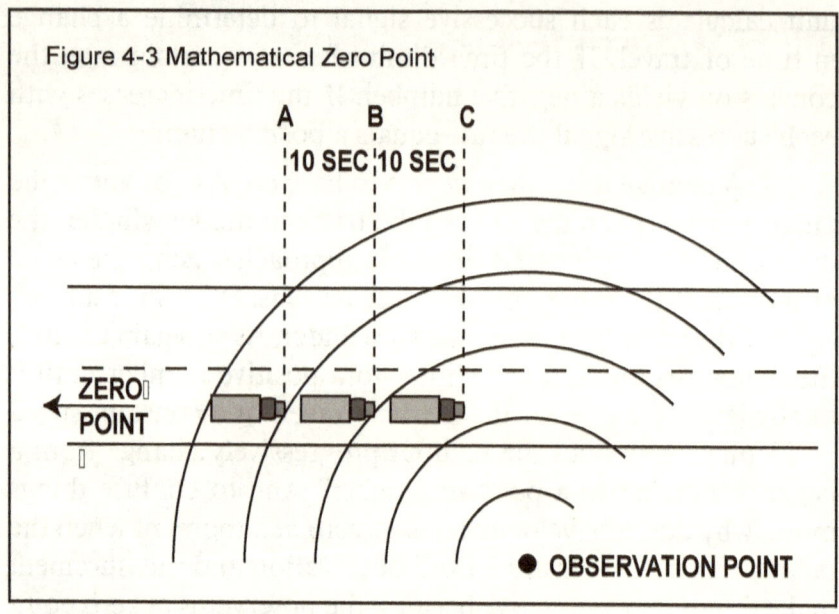

Figure 4-3 Mathematical Zero Point

The classical method of measuring velocity, assuming no error in measurement, requires two people to synchronize watches and walk in opposite directions upon the highway with one observer at the zero point of origin and another at the finish line. As the moving vehicle approaches the first observer at zero point, the time is noted the instant the vehicle passes. After this passing, the zero point of a mathematical observer marks the spot on the highway and attempts to set a tape measure to that point and walk the total distance traveled to the finish line observer, who in the mean time notes the exact time the vehicle passes the finish point. The two observers compare notes upon meeting and note that both watches appear identical. By dividing the distance between the zero point and the finish line by the "constant" of time marked down between the two observers, the velocity in measured mathematical terms is realized. The essence of the whole measuring process places a single zero point both for the distance traveled and the time elapsed. The zero point lies on the coordinate of travel, does not move, and is not on the sidelines.

An observer, standing on the sidelines, cannot assume that what he or she observes equals what the equation calculates

The Geometry of Mathematics | 78

any more than the radar unit would remain constant from any position eccept the point of origin. Mathematics demands a totally different conclusion than general sideline observation. All the rhetoric of relativity and the imaginative thought experiments cast a meaningless frame once the equation uses velocity as a vector. Mathematics has the observational mind's eye geometry only at zero. As soon as the observer leaves the zero point, the observation does not apply and the geometry of mathematics is meaningless.

5. Much About Nothing

> Energy and time exist, but are not objective. If you want to drive yourself crazy, try to think of time as objective. We say, "Give me time," but the transaction does not involve handing anything over.
>
> -Arthur M. Young[1]

Breaking the Time Barrier

With the advent of supersonic aircraft it has become quite familiar to brake through the sound barrier. This has been a very interesting phenomenon indeed and more the stranger when expressed as breaking the time barrier. Consider a supersonic jet on a straight level course traveling at a speed in excess of the velocity of sound. On approach the jet will diminish its observable distance from a listener, standing directly ahead of the jet, more rapidly than a sound wave is received because light travels faster than sound. It works on the same principle as holiday fireworks. We see the burst of colors with a delayed bang that follows. Therefore, the listener would receive the sound of a jet from a later moment during the observable flight of the jet. Because the jet travels faster than sound we can actually hear a later sound emission earlier than emitted at a further distance. Likewise, if the jet were traveling toward us at the speed of sound we would here all the sounds emitted over time to arrive at the same time as one high pitched bang.

If the supersonic jet travels at twice the speed of sound and sends out an audible sound at precisely one-second intervals and this according to a clock on board the jet, a recording device on the ground ahead of the jet would receive the later sounds before the former. If each sound pulse could be identified as to the time it left the jet by some special code, and if each successive sound shows up on a digital readout by some means of conversion, time would audibly appear to flow backwards. The jet would send each successive pulse as five; six, seven and eight, but the recorder would receive the very same pulses as eight, seven, six and five.

To make this point about time more relative to the discussion, suppose a jet flies away from a listening point at one half the speed of sound. At three-second intervals the jet returns a sound pulse. Each pulse may last for a portion of a second, but according to a very accurate clock traveling with the jet, each sound pulse begins precisely every three seconds. For the velocity of the jet to be one half the speed of sound relative to the listening devise, the recorder must be positioned in a line directly behind the jet at some arbitrary zero point of measurement or the mathematics of velocity relative to this point means nothing. By comparing each pulse received at the ground station with a stationary clock next to the listening device, each signal will arrive at 4.5-second intervals and not 3. The listener in this case would conclude that the timepiece traveling with the jet audibly appears to be moving slower.

By using the proper transformation equation, it can be determined precisely how slow time appears to pass on the moving jet relative to the ground listener. If we take the time of the stationary clock and divide it by one minus the ratio of the positive directional velocity of the jet and the negative directional velocity of the sound, we can calculate the time the jet clock appears to the listener at the zero point of the equation. The following equation illustrates:

$$\text{JET TIME} = \frac{\text{Ground Time}}{1 - \frac{+ \text{ jet velocity}}{- \text{ sound velocity}}}$$

The Einstein Illusion

As determined at the listening point, the jet time would appear to move slower than the frequency of stationary time. If ground time showed 9 seconds, then jet time would audibly appear to the ground listener as 6 seconds:

$$\text{JET TIME} = \frac{9 \text{ seconds}}{1 - (-.5)} = 6 \text{ seconds}$$

Since the jet velocity extends in the positive direction, the quotient of the ratio between the sound traveling in the negative direction and the jet traveling in the positive direction reduces the equity of one, thus yielding 6 seconds for apparent jet time. Nine seconds over one would give the jet time if it had zero velocity.

The direct proportion of the magnitude of time to the distance traveled can also be represented geometrically as in **Figure 5-1**. This sort of geometry will display real spatial dimensions horizontally but represents time vertically. This apparent right angle observation simply shows that time seems apparent and does not imply the actual slowing of the clock. As the jet travels at one half the speed of sound, the representation

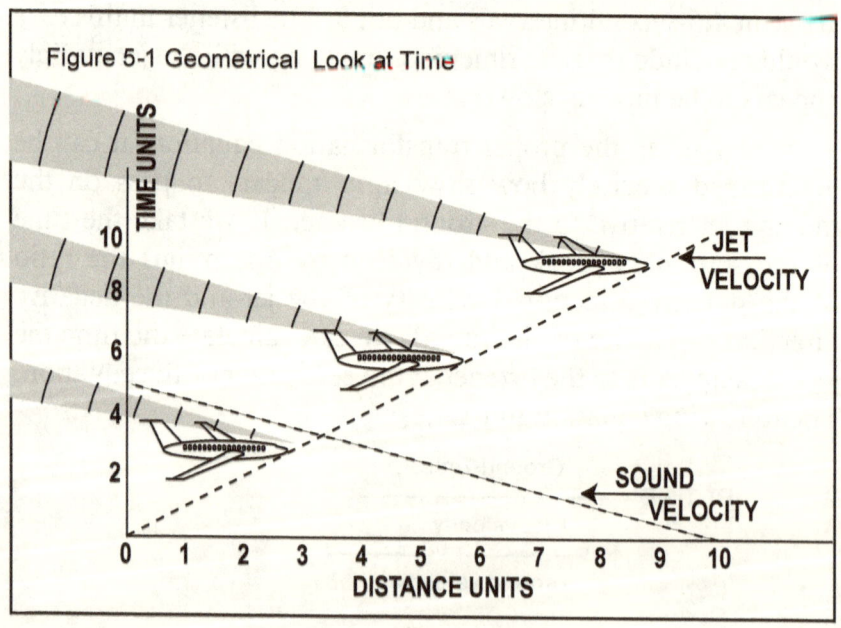

Figure 5-1 Geometrical Look at Time

of the velocity of the jet comes from an angle twice the angle represented for sound. The angle is not really directional. It only converts a magnitude to a perpendicular coordinate in order to determine a scalar intersection point.

Time gets its dimensional or symbolic property when represented at right angles to the direction of travel. But time has no direction. Time obviously does not move upward. Time has no more spatial reality than the gross national product on an economist's chart. Time manifests a mathematical relationship of magnitude and not dimension. Velocity represents an angle proportional to the time and distance. The whole analogy takes a series of photographs from the same 90 degree sideline position and stacks them one on top of another. Each photograph, taken at three-second intervals, displays the jet advancing to the right with an equal distance between each photograph. Because of the proportional magnitude between time and distance traveled, assuming a constant velocity, each picture could be placed at equal distances vertically, one photo on top of the other. We thus can trace a diagonal over the photographs as shown in **Figure 5-2**. Each diagonal represents either the velocity of the jet or the reverse direction of sound.

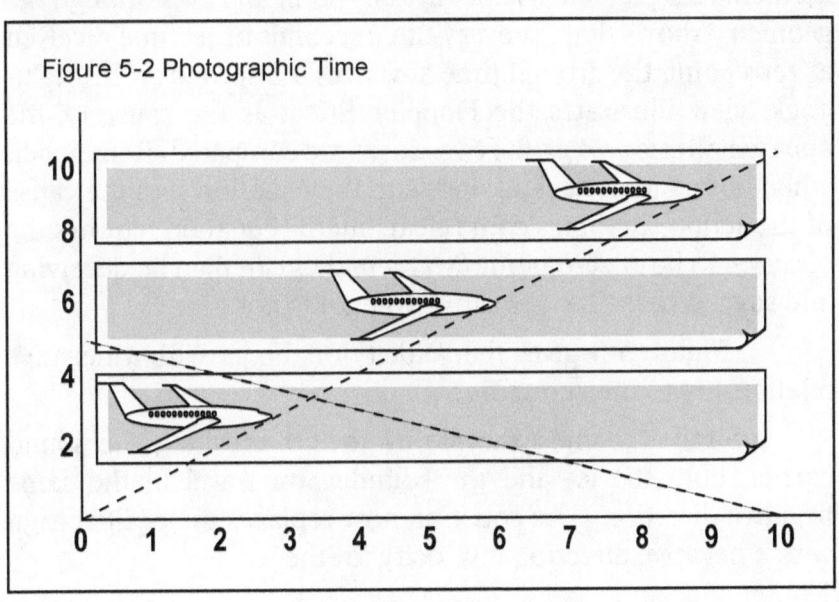

The Einstein Illusion

Once each pulse leaves the jet, it travels at a certain velocity relative to the surrounding air and since the sound moves at twice the velocity of the jet but in opposite directions relative to the zero point of the equation, the greater angle represents the velocity of sound. At three-second intervals of jet time a pulse of sound oscillates toward the zero point. By drawing pulse lines parallel to the sound velocity, the time of arrival at zero point or the recording device can be graphically determined. For every 4.5 seconds of ground time, jet time measures only 3 seconds. But, from the sideline photographs, the clock on the jet and the clock at zero point, if both could be photographed, would show that each clock shows agreement. Observation at right angles determines the agreement of both clocks but the mathematics from the zero point of velocity illustrates a difference. Since both clocks sit equal distance from the sideline camera, the speed of light will have no effect in separating the observed time from each clock. Even if the time arrived audibly to an instrument near the camera, no difference would be measured.

The illustration emphasizes that the mathematical equation and the geometry of the graph calculate the same thing, but the visual aspects of this graph yield the truth. The mathematics puts the listener at zero point and even though the geometry shows that for every three seconds of jet time received at zero point, the ground time advances 4.5 seconds. This right-angle view illustrates the Doppler Effect as the cause of the apparent illusion when the two clocks are compared. Remember, when observation at right angels to the equation sees the cause of the action, all things come clear, but the equation requires the observer to be at zero point. Mathematics can then be deceiving and suggest that time is relative when it is not

Figure 5-3 gives the velocity of the jet with a negative relationship to the zero point.

Similar to the example of the jet breaking the sound barrier, both the jet and the sound wave travel in the same negative direction. The equation now replaces the positive sign with a negative directional velocity for the jet.

Much About Nothing | **84**

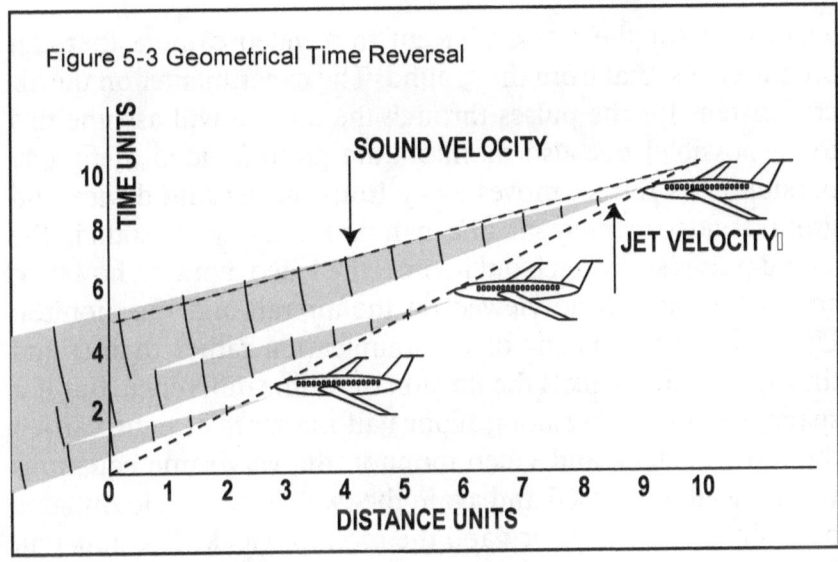

Figure 5-3 Geometrical Time Reversal

$$\text{JET TIME} = \frac{4.5 \text{ seconds ground time}}{1 - \dfrac{-\text{ jet velocity}}{-\text{ sound velocity}}} = 9$$

Or in the reduced form:

$$\text{JET TIME} = \frac{4.5 \text{ seconds}}{1 - (+.5)} = 9 \text{ seconds}$$

For the negative velocity to have any meaning graphically, the line representing the velocity of the jet must pass through the zero point. Still represented vertically, time matches the equation by being inverted. When compared with the velocity of sound, the effect that velocity has on apparent time easily demonstrates that the equations can also be used if light governs the means of communication. The similarities in the above equations assume that light as well as sound travel through a medium, and that the zero point remains at rest relative to the same medium. If the zero point were to move within a medium during travel, the equation transformations would become more complicated.

We can repeat the same situation differently. We shall assume a video monitor and receiver aboard the supersonic jet to monitor the clock adjacent to a video camera that can broadcast a signal from the ground. The experimenter on the air craft listens for the pulses through the air (we will assume this to be possible) but also monitors the ground video (definitely possible). As the jet moves away from the sensing device and ground video camera at one half the velocity of sound, the sound pulses still stretch out in time; the video monitor, however, shows the same time viewed on the aircraft and the monitor. Using light as a means of communication rather than sound through the air cancels the illusion of a time difference. But if a spaceship, advancing along at one half the speed of light, carried the moving clock and video monitor, the very same equations would need to be used and again the two clocks would disagree between the aircraft clock and the monitor clock. The apparent breaking of the time barrier is clearly understood. As in sound we can also illustrate the breaking of the light barrier.

The purpose of the forgoing analogies is to show how equations do not always predict observed reality when time dilation is considered. We should also understand that if communication were instantaneous in the case of light there would also be no difference. In talking about relativity the thought experiment observer is not always placed at the zero point, yet he is expected to observe things according to the equation as if he was at zero. When we observe velocity it is something different than what the mathematics of velocity depicts. Observation can be from any reference point, while mathematics is only from a fixed zero point. To say that we would observe something predicted by the equation is incorrect unless we stand at the zero point of every mathematical calculation.

Classical Illusions

An attempt can be made to set up a thought experiment that will demonstrate optically the classical theory of relativity from the points made thus far. This will help us develop a bit closer a better understanding of the Special Theory of Relativity.

Einstein never really explained the Special Theory other than writing the equation differently than in the manner explained thus far. We will come to that point, but first we must discuss classical relativity in a different way than Einstein. Although Einstein used classical relativity to introduce the concept, he never explained it as taught today. In order to eliminate concepts that deny comprehension, assume that the laboratory expands the size of a nuclear accelerator and the object of study can be measured by state of the art video equipment. The average person, perhaps, can understand the set up in **Figure 5-4**. Close to conceptual possibility, this example shows exactly how the equation interprets velocity in a proper transformation equation.

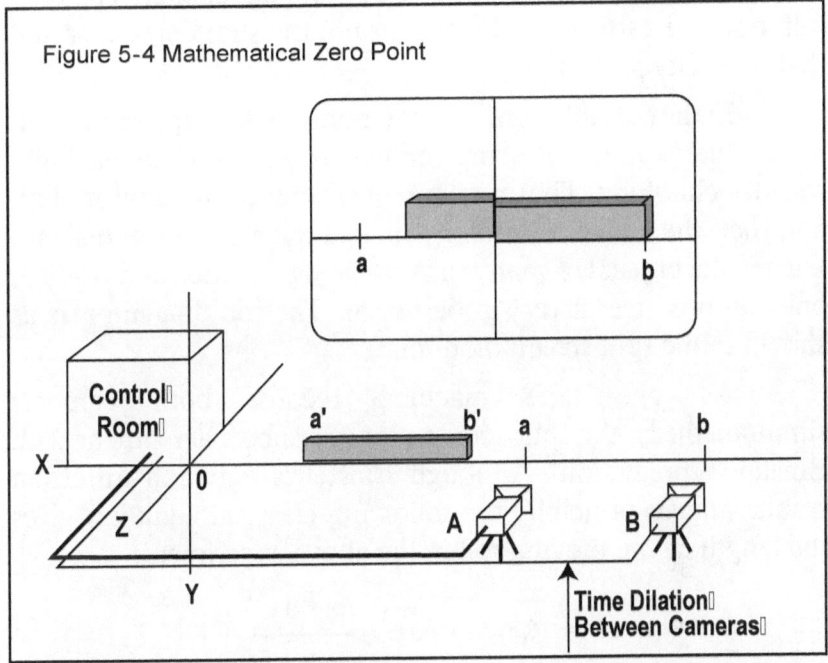

The x coordinate illustrates the track for which a moving rod travels. The length of the rod at rest equals the distance between points $a - b$ as marked on the track. Camera A and B stand so as to monitor these points when the rod passes. Positioned at the zero point and equal to the equation or definition of velocity, the control room defines velocity relative to this point. Any other position would give a false measurement of velocity. Even though video cameras are used, the control room equates the observer at the zero point of a mathematical positive velocity while the side-line observation is still illustrated by the cameras.

Overcoming the obvious difficulties, assume that the rod can be accelerated to a high enough velocity and that technology can effectively freeze frame the motion in the control room. By placing both a clock and a trip switch at zero point and also at the finish line at *b,* both switches will stop when the rod passes. When the beginning of the rod passes the zero point, it stops the first clock. When it passes point *b*, it stops the second clock. The elapsed time is determined by bringing the two clocks together before the experiment to synchronize and again after. The time difference gives us the velocity, but not the shortening of a rod that relativity predicts.

In the control room, several monitors and tape equipment replay the experiment using the electronics of assumed high-speed technology. The instant replay appears in slow motion and then the added technology freezes the picture for analysis. During the process of many runs, all being recorded and studied, one run produces a very good visual. The rod thus appears as shown in the split screen monitor.

As one tape machine records both cameras simultaneously, the split screen monitor shows agreement with Einstein's prediction? Although Einstein's equation differs in results and in principle, the following classical equation gives the length of the moving rod in the above experiment:

$$\text{MOVING ROD} = \text{ROD AT REST} \times \frac{1}{1 - \frac{+ \text{rod velocity}}{- \text{light velocity}}}$$

The moving Rod *a' b'* equals the rest space rod *a b* times one divided by one plus the ratio of the positive directional velocity of the rod divided by the negative directional velocity of light. The space *a b* equals the rod at rest, but in motion the moving rod *a' b'* equals the rod at rest multiplied by this classical transformation. The moving rod appears shorter than the rest length. Substitute quantities such as a two-meter rod moving at half the velocity of light and the following illustrates the point:

$$\text{MOVING ROD} = 2 \text{ METERS} \times \frac{1}{1 - (-.5)} = 1.3333 \text{ meters}$$

As shown in the split-screen monitor, the rod appears approximately two thirds the size of the space marked off as equal to the rod at rest. If the rod moved at the velocity of light, the equation would yield (for a two-meter rod at rest) a value of one meter for a moving rod. Even at rest the same equation would yield a value of the correct two meters.

Assuming light to be particles of information, the same foreshortening conclusion would result from classical mechanics. The foreshortening, obviously caused by the additional cable length between cameras illustrates that each camera differs in respect to time. It takes longer for the signal to reach the control room from camera *B* than from camera *A*. It must be understood that the velocity of both the rod and light must use the same zero point. In fact the transformation equation demands it. The signal travels at a negative velocity relative to the control room at the speed of light. At rest, the rod would show no difference, but the higher the velocity the greater the time difference. Look at the set up again in Figure 5-3. The moment the signal from camera *B* arrives at the camera *A*, the image of the two cameras will not be in sync. Camera *A* will have a later moment in time due to the time dilation. The signal from camera *A* would show the rod at a point past the mark. The older signal from camera *B* is due to the time it takes the signal from camera *A* to reach camera *B* so that the two signals will arrive simultaneously in the control room. However, they did not originate at the same time.

A right angle graphic close up of the moving rod in **Figure 5-5** will optically show that, when positioned precisely at the two meter position, a light photon particle reflects from point *b* and then travels in the opposite direction of the moving rod toward the observer at zero point of the measurement of velocity.

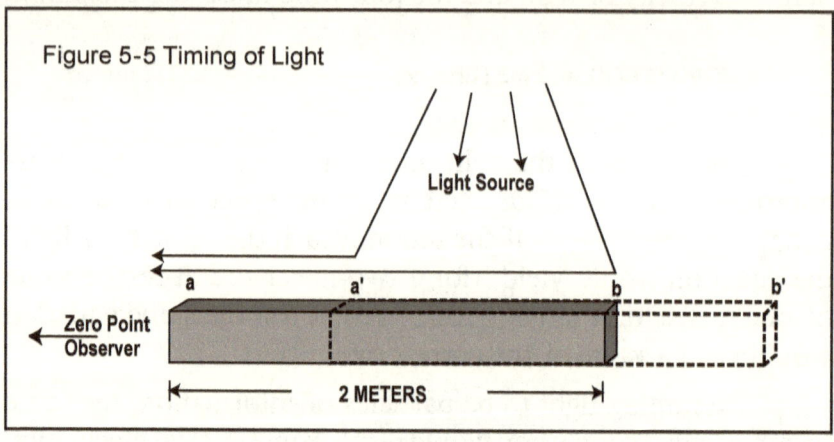

Figure 5-5 Timing of Light

This time period represents the distance electric current would travel in the cables between the cameras. Electric current, considered to have a similar velocity as light, illustrates the same zero point observation. When the photon particle reaches point *a'*, the rod has moved one-third its total length from *a* - *a'* due to the rod traveling at one half the velocity of light. At this precise time another photon particle reflects at point *a'* and travels with the first photon reflected from the opposite end originally at *b*. The two photons eventually reach the human eye or video monitor simultaneously. The two photon particles, however, did not leave the moving rod simultaneously. The distance between *a'* and *b*, obviously less than 2 meters, creates a mechanical illusion wherein the rod appears to be shortened. **Figure 5-6** demonstrates the matter graphically with time as a symbolic vertical.

As the light photon particle moves to the left two meters in .00000666 seconds (2/300,000 = .00000666), the rod will move one meter in the opposite direction during the same time (1/150,000 = .0000666). The intersection point lies at .666667 meters. Classical mechanics gives the two meter moving rod with an apparent length of 1.333 meters.

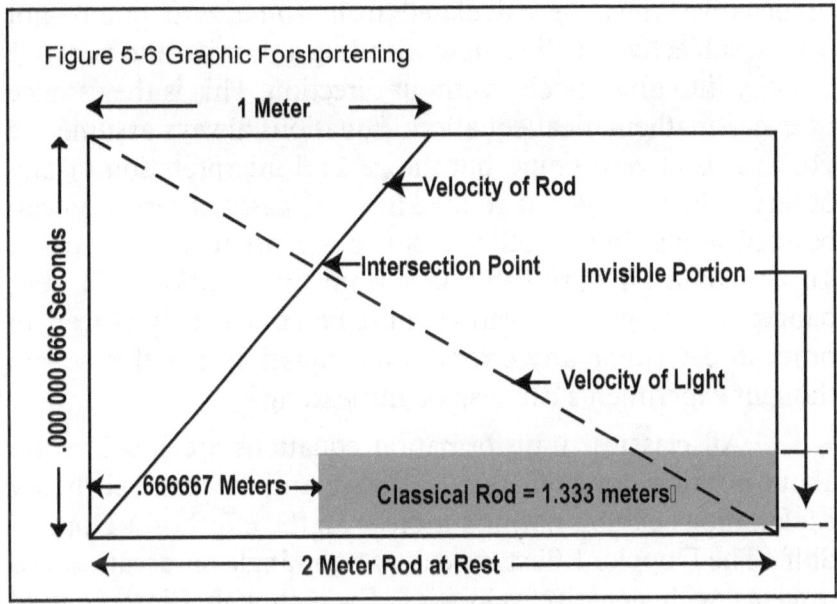

Figure 5-6 Graphic Forshortening

Move the control room to a different observation point; say at right angles to the direction of travel, and this right angle geometry will enhance the ability to understand the limitation of the mathematical equation from the zero point. Since both camera cables lie at an equal distance from the track and since both cameras stand an equal distance from the control room, the conclusion can be made that two photon particles would reflect from both point *a* and *b* positions simultaneously. Regardless of the speed of the moving rod it will appear the same in motion as at rest. The mathematics of the rod length or its velocity, however, means nothing from this observation point. When using the velocity of light in the equation, mathematics requires relativity measurement to come from the light traveling the length of the rod. Try to measure the length of the rod at right angles using the velocity of light. The true meaning of apparent relativity appears. The sideline cameras do not receive a single photon particle that has traversed the length from *a* - *b*. To use reflected light to measure something exemplifies one principle, but to use the *velocity* of light constitutes another factor. The equation of Einstein's Special Relativity assumes the constant velocity of light along the direction of travel, but velocity, whether positive or negative, must be calculated from some zero point—not from a subjective sideline view, blending a positive and negative velocity into one velocity without direction. This is the essence of every mathematical equation. Equations always assume the observer is at zero point, but the general interpretation of any observer does not visualize this. In some cases observation can be misleading, but in other cases, observation tells the truth. To conclude an observable fact from an equation is a very dangerous thing. Observation must be strategically placed in order to determine any errors in the equation. For this reason thought experiments can also be misleading.

All classical transformation equations are much to do about nothing, for both classical wave and mechanical theory explanations exhibit nothing more than the effect of a Doppler Shift. The Doppler Effect cancels at right-angle observation and reverses with negative velocities. Even though Einstein used the Doppler Shift to explain basic principles of relativity, the question arises, "What does Einstein's transformation equations predict in the Special Theory of Relativity?"

Einstein's Transformations

The medium, in the case of sound, constitutes the surrounding atmosphere and in the case of light a wave in a medium was also assumed. The above equations would not work if light or sound did not have a medium to propagate in. The equations would fail if the zero point, or observer, moved relative to the medium. Since this condition does not exist in real life we need to understand if it is possible to observe things symmetrically without a medium for light to propagate in. In the case of sound, consider a jet that travels at the speed of sound relative to the surrounding air. It would be impossible for the sound of the jet to leave the forward path of travel. **Figure 5-7** will illustrate.

At each position of the aircraft over a given distance a circle of sound is initiated at intervals from a center and

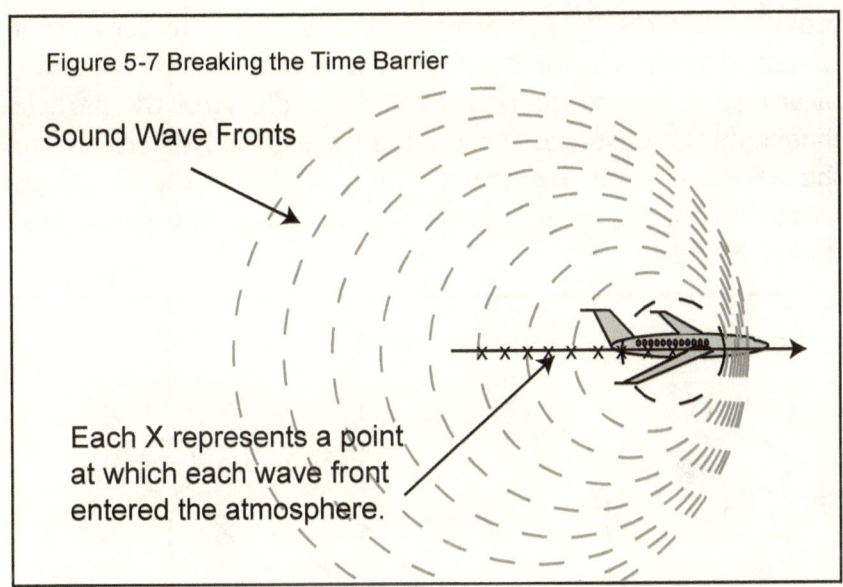

Figure 5-7 Breaking the Time Barrier
Sound Wave Fronts
Each X represents a point at which each wave front entered the atmosphere.

propagates outward. The sound builds up and creates a powerful force in the vicinity of the aircraft. Any pilot experiencing this phenomenon will plainly admit that the jet must travel slower or faster than the speed of sound to escape the vibrations of the sonic boom or wave-front build up. It is like emitting sound that never leaves the aircraft because both the aircraft and the sound travel in the same direction and at the same velocity. The surrounding air becomes the reference frame for the velocity of sound and not the source or the listener. This same conclusion resides in the classical wave theory of light. The measured velocity of light in this case will always be the same relative to the medium it travels in, no matter what the motion of the observer making the measurement, and no matter what the motion of the light source.

A natural opposing theory would describe the mechanical theory. Under this theory of light, the measured velocity will always be the same relative to the light source, no matter what the motion of the observer making the measurement. The mechanical theory provides no inhibiting medium. In both the classical wave theory and the mechanical theory the classical equations would yield similar results as the video monitor provided the results in the control room and the zero point remained at rest relative to a medium in the case of a wave theory, or at rest relative to the light source in the case of the particle theory. If the observer and the light source moved relative to the reference frame, distortion would result and the equations would no longer apply. **Figure 5-8** illustrates the mechanical theory of light.

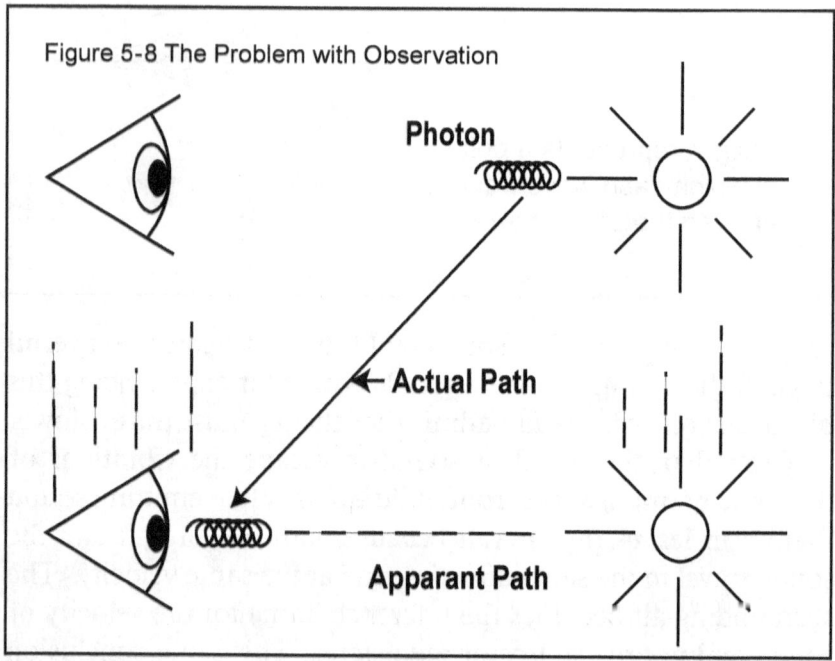

Figure 5-8 The Problem with Observation

The difference between the apparent and the actual path that light travels under the mechanical or particle theory, exhibits no apparent change to the observer as long as both observer and emitter are at a constant distance from one another. The mechanical theory treats light as bullet-like particles. A bullet, fired at an observer, would never reach the victim if the gun accelerates away from the observer at the velocity equal to the negative velocity of the bullet. On the other hand a particle-like bullet would arrive at its target at twice the velocity, if the one firing the gun would be moving toward the target at the normal velocity of a speeding bullet. There are obvious limitations to the mechanical theory. A rotating light source provides the same problem. It will no longer appear to have its actual circular orbit. The greater the orbital velocity, the greater the apparent orbit appears to the actual orbit. Distortion in the particle theory magnifies but remains somewhat symmetrical. Things are not as they seem.

Since the universe has pockets of galaxies and double stars that maintain a constant circular acceleration that does not show the distortion required by the particle analog of light, the scientist rejects the classical mechanical theory for light propagation. The only other alternative left is some kind of wave theory, but it cannot be where a light source or observer moves within a rigid frame, otherwise we again have a distortion. The medium must have some sort of dynamic properties that alter the velocity of light to a consistency surrounding large bodies at a given distance. It would also have to be a situation where the velocity decreases the closer it propagates near a planet or star and accelerates to some extent when a celestial body is far distant. This agrees with the General Theory of Relativity often referred to as space becomes denser near a large body thus causing light to slow and also curve. Writers use the expression "gravity becomes more dense." This author has changed that expression to say that the magnetic field becomes denser in the vicinity of a large body and not gravity.

Modern physics suggests that a gravitational field will curve light and slow it down, but as to the Special Theory of Relativity the new physics refuses a medium altogether. If the gravitational field can slow light and cause it to curve, why do we not attribute the same to the magnetic field? If this conclusion follows the inverse square law as will gravity, how do we not know that if a change in magnetic density can cause light to curve perhaps all things curve for the same reason? **Figure 5-9** will illustrate that the traditional lines of force for the electric field is the same for gravity. Both follow the inverse square law. Of course there is a difference in terms of distance and force.

In both situations the two fields are only conventional analogs. The assumption that the forces of the fields run in the

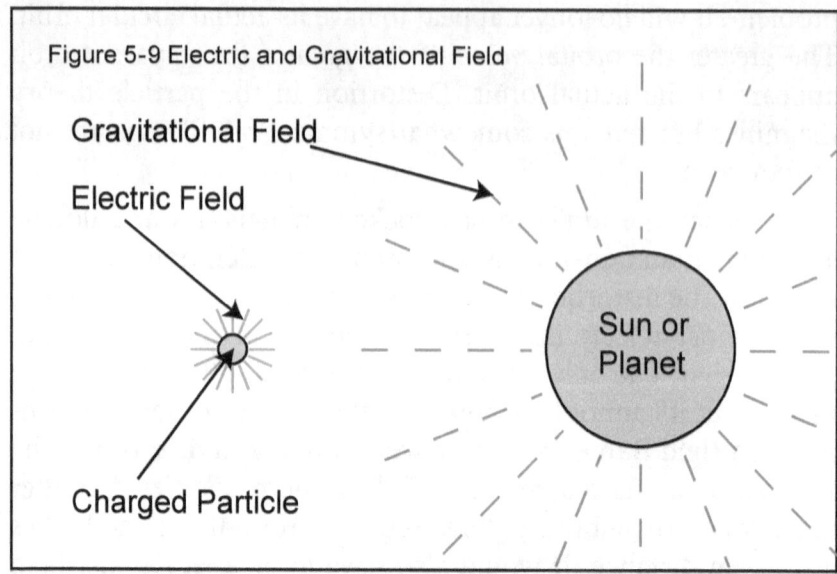

Figure 5-9 Electric and Gravitational Field

direction of the lines of force is somewhat foolish because a force of accretion is a classical assumption based upon old action at a distance theories. Instead of perpendicular lines of force think of a graduated decrease in magnetic density in both cases. It is not as if something reaches out and pulls, but that when electric properties, including light pass through the graduated magnetic density affect a directional change or curvature toward the body. The strength from one system or another is the degree of change in magnetic density. Putting aside charge for a later analysis, consider than the size determines the basic graduated density from one level to another. Even though gravity obeys the inverse square law on the earth as well as the sun consider than the graduated magnetic density diminishes faster per vertical distance from the surface of the sun than from the surface of the earth. Particles two would have special diminishing characteristics. The force is how the electric fields in light and matter react to this change of magnetic density.

Instead of perpendicular lines spreading apart like the rays of the sun why do we not say that the magnetic field is denser the closer it is to a particle planet or star. In this way we can have a field that circulates the particle or planet and returns.

The change in density of the magnetic field is what causes light to slow down and curve. What we have then is magnetic space everywhere to govern the velocity of light. No one seems to grasp this conclusion. They preferred empty space to magnetic space. They also prefer that energy such as gravitation and other forms are non-material and can be magically converted to material and back again. They also prefer that light is a form of this pure energy without substance. Who is in the dark ages? You decide. Is it modern physics or this Author? This author, Michael Faraday and theism are generally in one corner and James Maxwell, Modern Physics, and atheism seem to prefer the other. I ask you to hold argument on this point until this thesis is finished.

At the turn of the century, the need of a theory of light that would solve the dilemma of distortion seemed eminent. Einstein's contemporaries escaped the problem by distorting space and time. Einstein's Special Theory of Relativity was not a distortion, but an explanation of Maxwellian space. It seemed to solve the dilemma, but it created a different set of paradoxes. Those paradoxes were such that the need to change the meaning of the terms to fit an assumed observation of empty space and the constant velocity of light.

Relativity theory produced a wave-particle duality without any medium to reference the velocity of light. Relativity used similar mathematics of a medium at rest relative to every observer as if in their own universe. It was like each and every observer had a personal ridged medium to reference the nature of light rather than let light slow down or curve in a magnetic field. The classical equation, illustrated earlier, could work with a variable magnetic medium, but for every observer to have a privileged frame of reference was contradictory. It became essential over time to warp space and time so every observer would be justified in being at the center of the universe. Unlike the mechanical theory, the light source does not affect the velocity of Einstein's photon. The Special Theory puts forth the assumption that the velocity of light is constant. The premise

The Einstein Illusion

states: *The measured speed of light will always be the same to every observer anywhere in the universe, no matter what the motion of the observer making the measurement, and no matter what the motion of the light source.* In actuality this means that the measured speed of light will always be the same relative to the observer making the measurement, regardless of the observer's motion relative to the light source. In other words, under the Special Theory of Relativity the velocity of light always moves relative to the observer and not the medium or the source. In essence, each observer has a personal mechanical-like special grid that follows him or her and that determines the velocity of light. Since light sources move it became essential to warp space and time in order to make the new equation fit the assumptions.

Even though it was a contradiction to common sense, the Special Relativity Theory assumed that the observer is the reference frame. Without the observer being in control, particle light would be like a baseball-pitcher and catcher riding on opposite sides of a merry-go-round. If the baseball is aimed at the catcher, the catcher might catch the ball if the pitch is fast enough and the merry-go-round moves ever so slow. If, however, the merry-go-round moved very fast and the pitch was slow, one riding the marry-go-round would see the ball appear to curve in the direction the picture is moving and opposite in the direction of the receiving catcher. An outside observer would see that the ball travels a straight line. Actually the reverse would appear if the ball was light in a medium that was spinning with the marry-go-round—an outside observer would think the ball is curving and one riding with the pitcher and catcher would see a straight path. This is what should be expected in the case of light and thus a medium of sorts is needed, but it must be dynamic and not ridged to produce symmetrical results.

Prior to Einstein, hope rested in the theory of a continuous medium that filled the universe. Such an idea, from the nineteenth century, revealed the belief in the luminiferous aether as a medium for the propagation of light. This concept proved non-observable. Science, to this date, assumes only that which can be observed despite the apparent nature of observation often being at sidelines to the mathematics. Classical theories, however, insisted upon the existence of things not seen.

With great reluctance Michelson-Morley, in 1887, failed to show a difference in the velocity of light. The experiment was repeated in many conditions by Michelson, Morley and others over the years. The most interesting example expresses the measurement of light from a distant star. The comparison was between the measured velocities of light in the direction of the earth's rotation or orbit toward and away from a distant star. One measurement would be a positive velocity and the other a negative velocity toward the star made at different times of the day or year. The many attempts made over the years wanted to see a difference in the velocity of light relative to an assumed rigid reference frame of the luminiferous aether. The rotation or orbit should reveal a slight variation in the velocity of light. The experiments were difficult, but accurate. In every case the experiments yielded no apparent difference in the velocity of light under any circumstances. The differences measured were too small in comparison to the rotation or orbit of the earth. They never considered that the magnetic field could be rotating with the earth and/or its density was relatively constant on the surface.

Even though Morley tried on later occasions, every experiment failed. He remained convinced this was due to partial entrainment. What he meant was not clear, but it would be similar to the interaction of different fields. Why the physicist of the time insisted on a rigid medium seems very strange indeed. One probably thought that a variable medium would cause distortion or refraction. These directions or partial entrainment were completely dismissed in favor of empty space. Can one really conceive of empty space where one will not find at least a variable magnetic field density? Keep in mind that the observer is at rest relative to this rotating magnetic field or is at rest relative to a specific density. These explanations were never considered.

The Michelson-Morley experiment was of considerable importance. Responsible for the downfall of the aether or classical wave concept, this aether wind did not appear observable. While Michelson-Morley created the experiment to test the aether wind theory and thus determine the constant velocity of light as relative to the medium, Einstein reasoned through Maxwell's equations the constancy of light propagation to the observer as a basis of light's velocity. Coming from separate directions, the failure of one and the calculations of another would simply agree. It is the calculations in empty space relative to every observer that instigated a paradox that has never been explained but is accepted nonetheless. Was it a desire more for strangeness than a desire for understanding? Perhaps it was a psychological problem rather than a problem of intelligence. Perhaps that problem was beginning to maturate under the auspices of atheism. The direction of Modern Physics does suggest such.

The many attempts made to explain away the paradox and the negative results of the Michelson-Morley experiment fostered several proposals in order to avoid the paradox of there being a number of reference frames equal to the number of observers. George Francis Fitzgerald (1851-1901) proposed that all objects grew shorter in the direction of travel. Fitzgerald, like Michelson-Morley, looked for some absolute motion—something that the velocity of light could be referenced with. He reasoned that the object would shorten by the pressure against the aether wind. The amount of this foreshortening would increase as the velocity relative to the aether wind increased. This would account for the failure of the experiment.

Despite the fact that thousands of observations made over many months and despite the fact that Michelson-Morley announced in July, 1887 that the aether wind did not exist, Fitzgerald wanted to save the aether as a light medium for wave propagation. Because of the extensive mathematical development of wave mechanics, the difficulty for Fitzgerald and others to accept light as something other than the wave motion of aether was difficult.

Hendrik Antoon Lorentz (1853-1928) put into mathematical form the Fitzgerald contraction. Lorentz developed this transformation in the course of a mathematical study of electromagnetism and applied it to an attempt to explain the Michelson-Morley experiment.

<blockquote>Einstein's basic mathematics for his Special Theory of Relativity was the same as the Lorentz Transformation, but he had not heard of Lorentz's prior work.[2]</blockquote>

The Lorentz's Transformation follows in **Figure 5-10**:

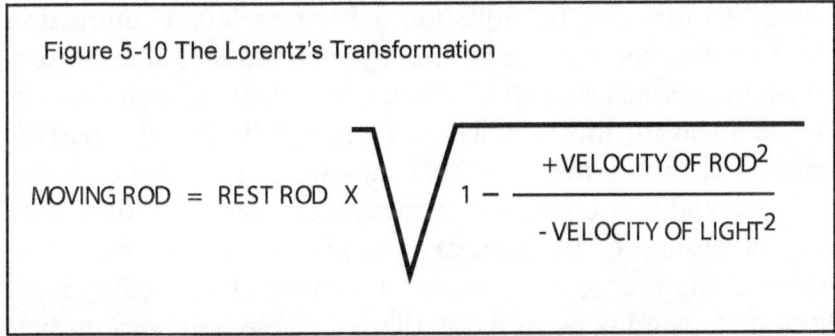

Figure 5-10 The Lorentz's Transformation

$$\text{MOVING ROD} = \text{REST ROD} \times \sqrt{1 - \frac{+\text{VELOCITY OF ROD}^2}{-\text{VELOCITY OF LIGHT}^2}}$$

The equation states that the moving rod shortens by an amount equal to the rest rod multiplied by the square root of the positive velocity of the rod squared divided by the square of the negative velocity of light. It is exactly equal in principle to the classical equation. The only difference is that Lorentz's equation does not produce a difference from a negative or positive direction. In other words, direction is pointless. It is like talking about velocity without a zero point or direction. It is mathematically meaningless when compared to the classical method. It should be noticed that the plus and minus signs preceding velocity as in the classical transformation equations do not affect the outcome as to direction because when you square the velocity, the outcome is the same whether in the positive or negative direction. The insertion of the signs only show the direction as demonstrated in the classical examples, but the result is not affected. Lorentz felt that the aether wind caused a pressure against the rod, thus shortening it to a degree relative to its velocity. The equation simply took direction out of the equation. We thus look at relativity as one would observe from the side lines, yet keeping the result as if from zero point. The equation is not a measurement as in classical relativity, but is manipulated to create a solution that would agree with observation—the same result for all observers.

As mentioned, Einstein developed the same equation as Lorentz, but from a different argument. Einstein actually used the Lorentz transformation in his own explanations. The Lorentz transformation tried to understand the Michelson-Morley experiment in a causal way (the pressure against the either wind). To Einstein, the equation did not explain a contraction due to the aether, but due to velocity itself would a rod shorten. Einstein assumed that all observers, regardless of their motion, would measure the velocity of light equally. Every observer could make the same claim that they were at the center of the universe and if you preferred a medium this aether moved with the one observing. In either case, all power was retained by the observer in order to make reality symmetrical, but reality itself became a paradox by thinking that every observer was in their own universe. The atheist most certainly delighted in this.

Later, Einstein used the Lorentz transformation to explain Special Relativity. In doing so, did he feel that the transformation reduces the actual spacial length of a high velocity moving object, or did he mean that the Lorentz transformation predicts an apparent observational event as in classical relativity? In another way, was Einstein concerned with observable reality as an objective event? He maintained that:

> It is neither the point in space, nor the instant in time, at which something happens that, has physical reality, but only the event itself.[3]

It is almost as if there is no past or future determination. Was the event itself, here referred to, nothing but observation itself? If so, the observed event had nothing to do with the point in time or the place of happening, it was the observation that surfaced as the only physical reality. In this way Einstein gives objectivity to observation and puts the actual point in space/time as happening within another observer's space/time. Essentially

Einstein might have admitted that time exists not as a separate dimension, but only as a derivative of space. Consider that Einstein admitted:

> The non-divisibility of the four-dimensional continuum of events does not at all, however, involve the equivalence of the space co-ordinates with the time co-ordinate. On the contrary, we must remember that the time co-ordinate is defined physically wholly differently from the space co-ordinates.[4]

Regardless of how one interprets Einstein, relativity seemed to explain the negative results of the Michelson-Morley experiment. Relativity appeared to destroy all absoluteness of space and time. Einstein may have had a clear and accurate picture, and he knew his equations were right in that they agreed with a new concept of observation, but his terms were a little confusing and he had no physical explanation. The observer became the culprit even if the aether was retained. Einstein seemed to reject the aether wind.

Although the Fitzgerald contraction of objects moving against the non-observable aether wind has been rejected along with the aether wind itself, consider that Einstein's contraction may not be due to the velocity relative to the observer. If not the observer, or the aether wind, or the source itself, what constitutes the reference frame for the velocity of light? The answer is simple if the modern physicist will receive it: ***The field density of magnetic space***. This is the author's conclusion based upon the ability to make all observations agree and also remove every contradiction. This solves both the paradox of no measurable medium and Einstein's observer paradox. Reason can then say, "Light does not travel relative to the observer's space/time. Light's initial velocity starts with an emission relative to the objective source and accelerates or decelerates relative to the surrounding magnetic field density generated by the ponderable objective matter. A change in the magnetic field density would produce a corresponding change in the velocity of light."

Despite the abolition of the aether concept, Einstein assumed certain classical mechanics of light in a systematic observational form. The error in the Special Theory of Relativity attempted to fit observed reality to agree with physical reality. It becomes apparent to the logical thinker that the observer cannot have a privileged reference frame. Unless by coincidence all experiments made to date measured the scientist observer at a fixed reference relative to say a certain magnetic field density. Until the same experiments can be made, say in a moving space station traveling through a magnetic field, we will not know what affect the density or motion of the magnetic field on the surface of the earth has on the velocity of light. A vacuum on the surface of the earth does not equal a vacuum far distant in space or on the surface of the sun. You can remove all electric matter within the confines of a vacuum, but you cannot remove the magnetic field and its corresponding density relative to the size of the planet in the vicinity of the experiment—vacuum or not. The typical response to this is that the magnetic field is not objective or particular in nature. This is perhaps the major error of modern physics. It cannot be proven one way or another, but Michael Faraday thought the magnetic field to be particular in nature and thus density is appropriately acceptable.

The observer cannot govern the velocity of light. Only the surrounding field controls all observed experiments. Did Einstein or any simple-minded theorist ever consider that the velocity of light was not a wave in a medium, but simply an electric field spiraling relative to the magnetic density in free space? The denser the magnetic field, the slower the velocity of light or the slower the linear direction of the electric field would be even though the angular spiral would have the same velocity at all times. Did the scientist ever consider that the magnetic field density remains relatively constant on the surface of the earth, thus causing the Michelson-Morley experiment to produce a negative result regardless of rotation or orbit?

Instead of strange distortions of reason, Both Lorentz and Einstein's transformation equations could also predict the magnetic field density of free space and thereby return space and time to conservation and adjust for any observed illusions by assuming a variable velocity of light in a variable magnetic medium. There is no other answer and to assume a constant velocity of light everywhere in the universe even contradicts General Relativity, light refraction, medium transfer, and common sense. Perhaps there never was an aether wind but that does not mean that the magnetic field was not the variable medium needed to smooth out the illusion. The variable in relativity equations should have been the velocity of light and not space and time.

The theory thus suggested earlier and will be considered from this point on is that everything is an electromagnetic relation including gravity. This theory treats both light and matter as one and the same thing other than within matter the eclectic field's orbit is a somewhat stationary form while in light the electric field spirals through space. The cause of the change is in the nature of the magnetic field in the immediate vicinity of an electron or other massive particle.

6. The Einstein Illusion

> People have occasionally been baffled and frightened by this use of "four dimensions" and have thought that in some mysterious way physicists or mathematicians can imagine four dimensions. Nothing could be farther from the truth.
>
> -Hermann Bondi[1]

The Doppler Paradox

The Doppler Effect, named after Christian Johann Doppler (1803-1853), says: If a sound source approaches, its speed causes the pulses of sound to strike our ears at a faster rate than they would if the source stood at rest. In other words, the sound increases in frequency on approach and decreases in frequency when moving away. It is believed that the same Doppler Effect pertains also to light and would manifest the same results if light were a wave in a medium. Since there is no medium to reference light the Doppler Effect may or may not apply to an electron firing photons when it moves toward or away from the observer. Doubt express here is an honest expression because the Special Relativity Theory does not provide for a change in the velocity of light regardless of the motion of the observer relative to a moving source. Modern physics does however insist on a change in color.

If the velocity of the fired photons is constant relative to the observer the velocity as the eye receives them would manifest the same color? Modern Physics simply does not equate light to be a particle of mass even though the electron looses mass every time it fires a photon. Without being a wave in a medium or a particle moving relative to the source or observer, how can you produce a Doppler Effect? A change in color will not do without an explanation. If we do see a color shift in light as the light source moves, how do we know the Doppler Effect to be the cause? If we do not know the structure of light, how do we know what color is? When we study photons as particles we conclude that the smaller they are the more red they appear. When we study light as a wave we conclude that the longer the distance from crest to crest the redder the light appears. If both particle and wave are of a certain color upon emission and the motion of the source or observer does not affect the velocity of light (as assumed in the Special Relativity Principle), how can you reconcile the Doppler Effect without destroying the conservation of energy and momentum and still save the Special Relativity Principle?

One must ask if the distance between photons affects the color or if the size of the photon determines what color is manifest? It would at least make sense that the emission of each photon would yield some space. Changing the frequency of emission should not change the size of the photon if the cause of emission is the same for each photon. If the distance between each photon determines the color then all photons would seem to be of the same size. If the size of each photon determines the color, then the Doppler Effect does not apply unless the photon changes its length due to the observer's relative motion. The only other explanation of color change is the reduction or increase of momentum of the photon relative to the observer. In this case the size of the photon does not change; only the impact upon the eye changes the color as the observer moves toward or away from the photon. This treats the photon as a particle mass and contradicts observation of rotating objects.

According to Einstein's Photoelectric Effect, a photon has momentum. Momentum would imply mass or some other density principle. As an observer moves away from a light source the impact would have to be reduced relative to the observer and this contradicts the Special Relativity Theory. Also if you reduce momentum you must of necessity reduce either apparent mass or velocity. An apparent increase in size would then need to be made as the observer moves toward the source. This would be necessary in order to produce a Doppler Effect. Wave mechanics does not deal with the Doppler Effect in this way. It describes a wave length and not a particular mass. The problem with wavelength is that a medium is required to govern the apparent velocity relative to the observer and light source. Relativity again concludes there is no medium.

What is really fathomed in physics today is the interaction of an electric and magnetic field. Thus if a light source moves away the interaction is stretched out and there is less energy or color to be measured. If the interaction is expanded then we must ask by what system is the magnetic field moving with. If the electric field is initiated by the moving source, is the magnetic field relative to this source or to the observer? Reason would conclude that the magnetic field has no relation to either, but must be relative to the center of a particular gravitational mass. If this be the case then the Doppler Effect is not relative to the observer or source, but to the center of a particular mass or system. If the observer resides at relative rest to this mass, such as the earth, the magnetic field would be relative to the observer. If an observer were to pass through a magnetic field then light coming toward the observer would shift to the blue and the same light moving away would shift to the red. It is not the relative velocity of the observer and light source to each other that would determine the Doppler Shift, but the relative velocity of either to the magnetic field. You cannot conclude that the magnetic field is privileged to either the source or the observer unless one or the other is as rest relative to the gravitational system. If the magnetic field is dynamic, which makes more sense, then we need to study stars moving away and toward in the North-South direction of the earth's magnetic field and compare with an East-West direction. There should be a difference if the earth's magnetic field is dynamic and moves in a North-South direction but remains relatively at rest in an East-West direction.

We can continue this argumentation and may be able to reconcile the Special Relativity Theory with the Doppler Effect. The two are a paradox to each other until the nature of the dynamics of the magnetic field is considered rather than empty space. I do not wish to challenge the relationship of color and energy; I only wish to establish a reason for the conclusion suggested by the Special Relativity Theory when the observer has always been placed in an East-West direction in the study of an either wind. This puts the magnetic field at rest relative to the observer. Concluding that the observer has a privileged frame of reference is in error when in reality it is the magnetic field that is privileged. If an electron moving through a magnetic field fires a photon then the magnetic field determines the velocity and not the observer. Likewise if an observer moves through a magnetic field it is still the magnetic field that determines the velocity of light. We thus have a Doppler Effect only if the observer or source is moving through a magnetic field. The magnetic field becomes the medium for propagation. This does not mean wave propagation of the medium, although it should be considered. We can take a particular electric field as a spiral through a magnetic field at relative rest and product the same results as a electric current in a wire producing a circular motion of magnetic space surrounding the wire. The principle is similar. This suggests particular characteristics of an electric field and particular characteristics of the magnetic field. This was suggested by Faraday and makes perfectly good sense. It this way we have both particle and wave characteristics in the relation. In this way a moving magnetic field relative to the observer or source would simply enhance the Doppler Effect. This is the only explanation that can reconcile the Special Relativity Principle and the Doppler Effect where in both will depend upon the same magnetic space as a reference medium. We are then left to conclude that all matter is a compilation of these two physical fields.

I wish to establish that there is reason for a color shift and it is not the relative velocity of light as it is in sound. Color is due to the relative mass density of the electric field. This gives mass to a photon and agrees with every particle experiment. Light is literally a chip off the electron. The velocity of light then is due to the relative density of the magnetic field. This gives wave properties to light as it spirals through magnetic space. Since the density of a photon would yield color, how do we justify the Doppler Effect in terms of the wave phenomena? I do not think you can without a better understanding of magnetic space and how the electric field responds. I will indicate in a later chapter that a particular photon is released due to the magnetic density permeating the electron or proton. The size of the photon would then differ if the electron is moving away or toward an observer, but the observer has nothing to do with this phenomenon.

I will continue this discussion when we take up the red shift. At this point it is more important to reconcile the Special Relativity Principle with the Doppler Effect. This means that you cannot treat light as a wave without a medium. I have suggested that the magnetic field is the medium, but not in the way that air and water are mediums. The principle is different. At this point we can only say that the Doppler Effect is due to a source or observer moving in the medium. This is simplistic, but will have to do until we can better understand the nature of light, preserve both the particle and wave properties of light and continue to remove the many contradictions and paradoxes in modern physics.

In review, let us suppose the Doppler Effect could explain the famous twin paradox in relativity. After synchronizing watches, one twin gets into a spaceship and makes a high velocity trip. When he returns, he compares his watch with the twin left behind. According to the Special Relativity Theory the traveling twin will show an earlier time and therefore appear younger. With velocities close to that of light, the "time dilation" should be large. This, in theory, makes possible ultimate visits to other stars in the galaxy and beyond.[2] Just as the zero point inhibits an apparent effect on the mathematics of classical relativity, should it not also have a limiting effect on Einstein's Special Theory? Neglected by writers, the twin paradox does include both a positive velocity and a returning negative one. This limitation hints to the inclusion of some sort of reversal if the relativist insists that the Special Theory is created by the relative motion of the observer. Who is to know whether it is the twin traveling in the space ship that moves or it is the observer. Either way the equation works. If the velocity is not reversed in the equation when the two come together, the cause is not the relative position of either twin but relative to some unknown conditions within space surrounding either the moving or stationery twin.

It is the Doppler Effect that requires a reversal of velocity, otherwise the concept of frequency increase and decrease would not make sense. Thus a color change must incorporate this velocity reversal in order to justify an appropriate color shift. In the twin paradox we have something different. If the Special Theory does not require a reversal, how can the Doppler Effect even exist with light? This alone says that the observer's frame of reference cannot be the answer. The twin paradox presentation does not use the zero point in mathematics properly. Perhaps the equation should be the reverse from the moving twin considering that the stationery twin is the one really moving. The twin on earth would appear younger to the twin in the rocket ship. The equation works both ways and thus a contradiction is observed. This voodoo of Special Relativity is the problem. If velocity itself is the cause and not the observer, then you need a reference frame other than a moving observer or source.

In order to save the Special Theory, the aging paradox cannot be due to the observer but rather to traveling in a magnetic medium in much the same way light slows down in a gravitational field. For just as a particle would contract at high velocities through a magnetic field, so also would light. Light does not change in mass. It simply collapses like a spring along its axis of travel and moves slower through an increase in density of the magnetic field. The change in velocity is due to a change in magnetic density which is much greater on the surface of the sun and earth than it is in outer space. What the Special Theory observer sees is correct, but it is not the act of observation that is the cause. It is the relative density of magnetic space surrounding the observer that determines the velocity of light and thus the contraction of objects. It is something on the order of a Doppler Effect, but I reserve to suggest the color shift is different as to cause. When the traveling twin travels at a high velocity through a magnetic field as does light, every electric field within the twin contracts when the axis is along the direction of travel. It is the magnetic field density that becomes the reference frame and not the observer.

On the surface of the earth the density would be relatively constant and thus explain the negative result of the Michelson–Morley experiment. The most important point is that it removes the observer as the cause and also removes any paradox. It also matches the Einstein and Lorenz contraction through magnetic space. The real result of the twin paradox may not be a time deferential as much as one of the twins contracting or slowing down metabolically in the direction of travel through magnetic space. A time deferential would only apply if the electric fields spiral slower in the direction of travel and thus affect the molecular structure of the moving twin in much the same way as it takes more time for light to oscillate in atomic clocks on the surface of the earth than at high altitudes where the magnetic density is far less.

According to modern physics, observation constitutes "fact," and since observation demonstrated that light had no medium to wave in and that light did not travel relative to its source, Einstein concluded that each and every observer, moving or not, would reference the same light with the same velocity. This statement implies a paradox and a contradiction, but if accepted, the observer in a real sort of way affects space and time, simply by the act of observing. The observer becomes the cause and the effect of seeing. Einstein did not consider the slowing of light due to magnetic pressure. Even under the general theory, the slowing of time was the culprit and not the velocity of light. To quickly solve for an isotropic universe, when the velocity of light was observed to move without a medium, modern physics formulated a theory that played havoc with direction by assuming the constant velocity of light to all observers. This new constant axiomatically defied reason and, even though the average person finds comfort in the word "constant," it seems incomprehensible for light to maintain a constant velocity without a source or medium to reference. To date, no one can explain it. How can light be bound by the same laws as other physical properties and still be dependent on the observer? This does not make any sense. Light cannot have a constant velocity any more than an electron. The mistake was to assume that nothing existed in a vacuum. This was modern relativity's grand mistake. The density of the magnetic field is obviously variable and different on the surface of a planet and outer space. Einstein came to his conclusion through Maxwell's equations. It should be said that the equations were based upon an electromagnetic relationship on the surface of the earth having a specific magnetic density. It cannot be assumed that conditions on the surface of the sun would be identical even in a vacuum. There is a difference between electric matter and the permeating magnetic field. You can remove one, but not the other. A vacuum is not a complete void.

If the whole of physics depends upon observation as a criterion for reality, then anything traveling faster than light, for all intents and purposes, does not exist. Observation is not actual reality but defines an apparent one. As T.S. Elliot might have put it, "Between the conception and the reality falls the shadow." Call it observation, the illusion lies between the subjective and the objective universe. Reason can explain the Doppler paradox of classical relativity by changing the observation point to the sidelines and thus cancel out the observed effect. Reason, also, ought to explain the observable Einstein Illusion by considering light traveling at a particular velocity relative to the magnetic space surrounding the observer and not to the observer directly.

Most simple-minded folks mistake the Doppler Effect for the Special Relativity Principle. But the Einstein contraction teaches real change; the objective shortening of moving rods (space) and a slowing of moving clocks (time) is real to the relativist. This describes the meaning of relativity. Real change, however, is not the problem when considering motion through a magnetic field. The illusion surfaces with a change in the meaning of the words "space" and "time." Just because a clock slows down in a dense magnetic field does not mean that time slows down. Just because a ruler contracts as it moves through a dense magnetic field does not change the meaning of space. By keeping the proper epistemology of the terms forces the mind to find a better answer. Changing the terms to fit observation simply created an illusion and a paradox that has kept us in darkness for over one hundred years.

Electric Contraction

The Lorentz equation came from a contraction theory, Einstein's equation came from Maxwell's field equations. Both will predict the observed effects of relativity, but the cause should never have been observer dependant. It has been explained that the Einstein effect, using the Lorentz transformation, differs from the classical transformation as follows:

$$\text{ROD IN MOTION} = \text{ROD AT REST} \times \underline{\text{Classical Transformation}}$$

$$\text{ROD IN MOTION} = \text{ROD AT REST} \times \cfrac{1}{1 - \cfrac{\text{Rod velocity}}{\text{Light velocity}}}$$

$$\text{ROD IN MOTION} = \text{ROD AT REST} \times \underline{\text{Lorentz Transformation}}$$

$$\text{MOVING ROD} = \text{REST ROD} \times \sqrt{1 - \cfrac{+\text{VELOCITY OF ROD}^2}{-\text{VELOCITY OF LIGHT}^2}}$$

The transformation equations under the classical wave effect, calculate both positive and negative results. Space and time appearances reverse with a negative velocity. But the Einstein effect plots a reciprocate quarter circle regardless of the direction. Assign a positive or negative sign to the rod velocity and also a sign for the direction of the velocity of light in the above Lorentz transformation equation, and the results will be the same. In other words, velocity has no direction when you square it. A positive multiplied by a positive equals a positive; likewise, a negative multiplied by a negative also yields a positive. **Figure 6-1** shows the Einstein curve from zero velocity to the velocity of light. In other words a two meter rod becomes zero in length at the velocity of light. The illustration compares other forms of observation in the confines of a Doppler Effect. Regardless of the negative or positive velocity the equation reveals the same Einstein curve. In the case of particles the Doppler Effect expresses a straight line either in the positive or negative direction. A wave in a medium has a slight curve. The illustration shows that the Einstein contraction is not due to the Doppler Effect because the positive and negative directions are unable to produce such an effect. If the Lorentz equation is true, you cannot suggest the observer has anything to do with

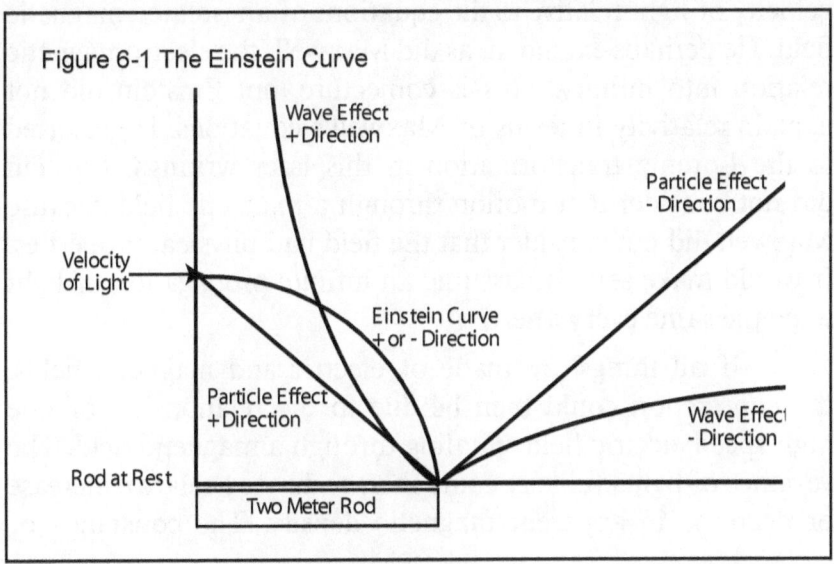

Figure 6-1 The Einstein Curve

The Einstein Illusion

the result. The Doppler Effect cannot have it both ways. If the observer is the cause then zero point must be in the equation requiring a positive and negative velocity relative to the observer or the source. If the equation cancels this observer dependence, then the velocity of light is not constant relative to the observer but rather constant relative to a specific magnetic density.

The Einstein effect plots the same curve in both the positive and negative direction and can be explained only in a causal way. Observation cannot play a part in the illusion unless the effect is reversible. Here lies the answer. Since the Einstein equation is not reversible, the effect is not reversible. If the effect on space is not reversible, it is not dependant upon the relative motion of the observer. For over eighty[3] years, this point has escaped the ever-popular special relativity enthusiast. It was not emphasized that observation, under any Doppler transformation, is reversible. If the equation is not reversible, the change is real and has nothing to do with the observer. The conclusion remains, the magnetic field density is the cause— case closed.

Maxwell's field equations may have implied a contraction of the electric field as it moves through a magnetic field at rest, but using those equations; Einstein mistakenly assumed the constant velocity of light relative to the equations of an isolated magnetic field. He perhaps extended, as did Maxwell, the electromagnetic relation into infinity. This is conjecture, but Einstein did not explain relativity in terms of Maxwell's equations. He resorted to the Lorentz transformation in this later writings. Einstein did not consider that motion through a magnetic field because Maxwell did not consider that the field had physical properties. It would make sense to assume an infinite propagation of light to be the same everywhere.

If all things are made of electric and magnetic fields, the contraction could then be due to the relationship of one high-speed electric field spiraling through a magnetic field. The variance of lights velocity could then be due to a relative increase or decrease in apparent magnetic density. The constancy of light must traverse a certain concentration of magnetic particle flux per second. If those magnetic particles become tighter, the velocity of light would be slower. Under this condition, the reversal would not be needed and Einstein's curve would work, but not because of the Doppler Effect. The observer would have nothing to do with observe results. This would make perfectly good sense if the observer was at rest to a particular magnetic field density. It is as if we measure magnetic particle density in space and not a mathematical form of empty space.

If the observer is at rest relative to a particular magnetic density; and if light slows down when passing through a dense magnetic field, then perhaps the aging process would also slow down while passing through the same field. Most certainly there would be a contraction. Keep in mind that the magnetic field density is relative, such that the faster the object moves through a field the greater the density appears to the object. Just as the electric field of light moves slower, so should all the electric fields of an aging twin move slower in the direction of their apparent spiral through a magnetic field. Absolute time would be constant, but measured time, such as that of a ticking clock, would slow, depending upon the change in magnetic field density that the measuring devise, whether ruler or clock, moves or resides in.

As relativity theory too quickly gave light a constant velocity relative to the observer without a medium, the challenging spiral-electric field theory gives light a constant velocity relative to the magnetic medium in which it propagates. The spiral electric field theory, pictured more clearly in later chapters, would also preserve, in certain ways, both the wave and particle characteristics of light. If others have also postulated an electromagnetic density relation, I apologize for assuming that it has not been done. The literature reaching one, who is a sage-distance from the thick of physics, represents only the popular and common textbook variety.[4]

Fundamentally, the only explanation for real change is that the electric particle field contracts against magnetic space at rest. If the observer is somewhat at rest relative to the magnetic field, it would naturally be assumed that the high-speed electric field would remain constant relative to the surrounding space of the observe regardless of the motion of the light source. For this reason, the observer was mistaken as the cause. To date, there has been many experiments moving a light source, but moving an observer at high speed through the earth's magnetic field has not been done. Perform the same Michelson-Morley experiment in a high-speed craft moving through a magnetic field, and the test ought to prove that the velocity of light will differ depending upon the direction through the magnetic field. If the Doppler Effect has anything to do with light, it is within this principle.

Consider doing an experiment in a north and south direction at a fixed point on the earth. Perhaps the magnetic density is greater near the magnetic north. All that the experiment has to do is test the velocity of light coming from both the north and the south. Because the magnetic field moves in only one direction we may be able to determine the velocity of the magnetic field. If the motion of the magnetic field is as slow as the earth moves east to west, there may not be a difference or it could be very small.

Like a spring, the faster the electric field spirals through a particular magnetic density, the greater the contraction. The observer assumes the velocity of light to be constant because the magnetic space surrounding the observation remains constant as to magnetic density. All experiments have been made relative to the earth's magnetic surface density and therefore relative to the observer. The velocity of light cannot be constant. Whether a rod runs positive or negative, the electric fields will contract as the tension builds up against magnetic space. It is called magnetic space, because the relativist may tolerate it better. Realistically, it should be called *magnetic particles in* space. This conclusion comes from Faraday, the model of which Maxwell built his equations and as Einstein continued. Maxwell denied the physical reality of magnetic space. I suspect that Einstein was more tolerable to magnetic and electric particles because he theorized the quantum particle characteristics of light.

Both Einstein's field theory of gravitation as well as Newton's action-at-a-distance gravitation suggests that particles, however small, cannot move in a path of travel without being affected by a distant mass. Now a particle curves in space either directly due to contact or indirectly through the magnetic field. The greater the mass density of a larger body yielding a greater variable magnetic density, the greater will be the smaller particle's curve. Is it the larger body that creates a diminishing magnetic field that encloses a smaller particle's path of travel with an uneven, side-to-side density, thus causing a change in direction? It is as if one side of the spiral contracts more than the other. Is this not really the same as the General Relativity Principle?

If you prefer action at a distance as if something reaches out from the center of a large body to the center of the small body you have a lot of explaining to do? Common sense would suggest that it is the magnetic space and its vertical variable density due to the larger mass that surrounds a small body and causing it to curve curvature toward the larger mass. Keep in mind that when a smaller particle's path of travel enters a greater magnetic field density and the density change does not have a side-to-side variation, is that the reason light will simply slow down without a curve? This would naturally be represented by light entering at right angles to a pane of glass—if not at right angles the magnetic surface density would vary across the width of a helical form of light causing it to refract. The degree of refraction would be due to the size of the electric spiral, thus we have color that is unrelated to the Classical Doppler Effect. The color shift to red or blue has to be determined by some relative motion, but that motion must be related to the apparent change in magnetic density and not to the motion of the observer.

There does not have to be a slowing of time or a contraction of space in the any relativity equation. Rather it is a slowing of light that will yield the same results. The slowing of the timepiece comes from the slowing of light in a denser field and not from the slowing of time itself. Light travels at a constant velocity relative to the field density the observer resides in and not relative to the observer. The linear reduction of an object's size occurs from the "apparent" increase in magnetic field density relative to the electric aspects of the object. This would be like a contraction of a spring without affecting the physical makeup of the metal itself. The greater the magnetic field density, the greater the electric spiral contracts in the direction of motion.

As Einstein's Special Theory of Relativity (inspired from Maxwell's equations of electromagnetism) can also imply that the compression of a rod occurs as it passes through a denser field relative to the moving rod and not to velocity relative to the observer. The observer is not the cause. If the relativist wants to keep his comfort zone and believe this then he also has to maintain an observer's zero point in the equations requiring a negative result with a negative velocity. A new kind of vector is still needed. Relativity equations need to calculate the tension of a magnetic field upon the electric. Perhaps this is precisely what relativity mathematics does, but we have failed to explain it as such. The tension that the magnetic field places upon the electric aspects of light determines both the velocity and the curvature of light.

A magnetic-dependant theory of relativity yields more understanding about Einstein's Special Theory of Relativity and his General Theory, and at the same time it develops a path of unification of electromagnetism and gravity. Magnetic relativity also explains all observed illusions including any paradoxes as causal to the magnetic field. The magic of the beholder evaporates as the velocity of light varies from one field density to another. The Lorentz-Fitzgerald contraction was right. The only adjustment is that the magnetic field becomes the medium and not the aether. Morley was right in using the work "entrainment" as a cause of the observed results. Both the Lorentz transformation and Einstein's equation say just this, but Maxwell's empty space must now be challenged.

A Bolt of Light

Modern quantum physics seems to conclude that light has specific units in that light can be broken up into smaller pieces or individual quanta. Despite this, the wave concept still permeates the electromagnetic concept of light. After demonstrating the non-existence of an ether wind from which light could wave in, the concept of light, as a wave, did not diminish, nor did the concept of light as a particle flourish. This was because the wave/particle duality was never resolved in a correct quantum electromagnetic model that maintains particle characteristics in terms of the two fields and also wave uniqueness in terms of a contracting spiral. The most common argument made is that light oscillates from an electric field to a magnetic field in free space. This does not explain quantum units. Even though the energy is broken up into individual quanta there is a failure to break up an electric field into the smallest constituents of mass. A wave length from crest to crest will not due. When refraction can be explained in terms of the size of a spiral, wave length is only a mathematical convention left over from classical wave mechanics rather than an actual depiction of light.

In 1902, Philipp Anton von Lenard observed that the energy of individual emitted electrons increased with the frequency of the light absorbed. This appeared to be at odds with James Clerk Maxwell's wave theory of light, which was thought to predict that the electron energy would be proportional to the intensity of the radiation. In 1905, Albert Einstein solved this apparent paradox by describing light as composed of discrete quanta, now called photons, rather than continuous waves. Based upon Max Planck's theory of black-body radiation, Einstein theorized that the energy in each quantum of light was equal to the frequency multiplied by a constant, later called Planck's constant. A photon above a threshold frequency has the required energy to eject a single electron, creating the observed effect. This discovery led to the quantum revolution in physics and earned Einstein the Nobel Prize in Physics in 1921.[5] The quantum revolution, however did not resolve the wave particle duality. We still talk of frequency of light in terms of wave lengths. This is because we treat light as energy separate from the mass it possesses. It is the mass upon impact that measures the energy. A wave front only gives us the energy in a medium. The electromagnetic reaction is not the same thing as classical wave mechanics and does not produce the same classical Doppler Effect.

The particle characteristics of light or the electric field still approaches justification in that Newton felt light to be corpuscles and Einstein's general theory of relativity says that a particle photon or electric field bundle suggests mass as it curves twice that of matter in a gravitational field. Light as a spiral field of particles ads to our understanding and tells us why planets orbit. Still the wave concept was preserved in the electro-magnetic relation because we fail to consider that the electric field is a string of photo-electric particle units. The curren electromagnetic theory does not resolve this issue unless in some way there is an attempt to reconcile string theory as what has been presented in this thesis.

The wave theory poses a problem because experiment denies the existence of a medium to carry the energy. The emission theory likewise poses its own problem. An orbiting light source would appear enlarged the farther its distance. The further a galaxy, the larger and more defuse its appearance. Both theories carry contradictions. To keep the universe in symmetrical order, common sense demands that light has a variable velocity relative to each change in the magnetic field density.

Light does become variable as it glances near the surface of the sun? Light also changes its direction in a gravitational field. This could be interpreted as a slowing of time in particular or a slowing of velocity in general. If we accept that the velocity of light diminishes as the field density increases, then light has a constant velocity as long as the magnetic field density is constant. The physicist says that the velocity of light does diminish when it passes through a denser medium. Is that any different than saying that light passes through a denser magnetic or traditionally called a gravitational field? This poses a new question, "How can atmospheric particles inhibit the velocity without collision?" There is no collision. Light either passes between particles at a slower pace, due to the magnetic density; light refracts around a particle, due to a close brush through an extreme density change near the surface of a particle; light reflects off the particle if it hits it; and finely light is absorbed by an atom becoming part of the total angular momentum. Particles do not slow particles other than in collisions. The magnetic field does this without a hitch.

Now if the variable magnetic field density is proportional to the mass surrounding the path of travel of light or any small particle, then the gravitational mass is not the direct cause of a gravitational pull—rather it is the variable magnetic field density. If light forms a spiral of photo-electric particles, natural assumptions can be made that those particles will react with a magnetic particle field as if at rest. The denser the magnetic field, the reaction time is condensed. Like walking upon a series of patio stones, the frequency of each step remains constant, but the walker will change only if the distance between stones varies. Likewise, if the distance between each magnetic particle condenses, a more frequent reaction to the electric spiral would result. Also, a slower velocity through a windowpane could be better explained by magnetic density rather than matter density.

Just as the field density of the magnetic lines of force differs through the universe, so also does the velocity of the photo-electric particle field. The foundation of the theories postulated in this book treat light not as a wave oscillation of two fields, but a bundle of photo-electric particles of light spiraling around a magnetic field density relative to some large planetary mass. The motion of the magnetic field may or may not have anything to do with the velocity of light unless the path of travel is parallel to the path of the magnetic field.

The nut and bolt analog will serve very well. With the nut serving as the photon particle bundle and the bolt serving the magnetic structure in free space, the threads will serve as the magnetic density. As the nut or electric particle bundle spirals with a constant angular velocity, the linear velocity changes according to the threaded density. When the bolt remains at rest relative to the observer, the nut passes through space to eventually hit the eye. If the nut remains at rest relative to the observer, the bolt would then split and circle in the familiar magnetic pattern and the nut would design the familiar structure of matter itself. Not so strange, all things are fields or bolts of light. If a nut was spinning across a bent bolt it would curve due to one side of the bolt having less threaded density than the other—fewer threads per inch because they are pulled apart. With matter the magnetic field circulates through an electric field at rest. With light the magnetic properties are more at rest and the photon bundle spirals through magnetic space. Why physics has lost sight of the simple is perplexing. It seems that complexity justifies the job, the grant and the mystery.

The Einstein Illusion is based upon the mathematical assumption that the universe is fixed relative to the observer. This would mean that each individual would have his or her own universe as if there were multiple zero points of measurements and multiple universes. The modern physicists need to search for a reason for this illusion rather than simply accept this paradox on blind faith.

Samuel Dael

7. Objectivity in Matter

> Physical concepts are free creations of the human mind, and are not, however it may seem, uniquely determined by the external world. In our endeavor to understand reality we are somewhat like a man trying to understand the mechanism of a closed watch. He sees the face and the moving hands, even hears its ticking, but he has no way of opening the case. If he is ingenious he may form some picture of a mechanism which could be responsible for all the things he observes, but he may never be quite sure his picture is the only one which could explain his observations. He will never be able to compare his picture with the real mechanism and he cannot even imagine the possibility of the meaning of such a comparison. But he certainly believes that, as his knowledge increases, his picture of reality will become simpler and simpler and will explain a wider and wider range of his sensuous impressions. He may also believe in the existence of the ideal limit of knowledge and that it is approached by the human mind. He may call this ideal limit the objective truth.
>
> -Albert Einstein & Leopold Infeld[1]

Objectively Speaking

Mass stands as a fundamental concept in physical science and can be treated as the most objective part of reality. It seems that we may never know this "ideal limit." It's like forever learning and never coming to a perfect knowledge. Many have invented philosophies that deny the very existence of objective reality because of the difficulty in laying hold of the smallest unit of tangible inertia. None the less, physicists define objective reality with proportions; three of which are:

$$\text{MASS} = \text{DENSITY} \times \text{VOLUME}$$

$$\text{MASS} = \frac{\text{MOMENTUM}}{\text{VELOCITY}}$$

$$\text{MASS} = \frac{\text{WEIGHT}}{\text{ACCELERATION}}$$

Space/time is needed to define velocity and acceleration, but space alone will define volume. On the other hand, *density*, *momentum* and *weight* can only be defined using a predetermined constant demonstrating the behavior of mass in a gravitational field. As mentioned earlier it takes some sort of directional component we call gravity in order to determine mass as a scalar quantity we consider as inertia. Gravity has produced this constant and is the variable used in each equation. For this reason the proportions of density, momentum and weight are produced using this directional constant determined by gravity. The inability to know the mass without the density demonstrates the same inability to know the density without the mass. The same also applies to weight, momentum, work, force, power and other mass related concepts. Each mass unit defines a proportional relationship between a determined gravitational constant in conjunction with a space/time proportion.

Experience attempts to generalize objective mass as an inertial substance that resists a change in rest or motion until acted upon. It is only because intelligence has been conditioned through experience that one concludes that inert matter appears to curve in gravitational space (magnetic space according to this thesis). Science defines the cause as a gravitational field. To date, there are accurate equations expressing this curvature but no one has postulated a direct cause, rather we are left with the classical assumption of action at a distance or the new concept of curved space. Both concepts pose certain paradoxes. In this, gravitation yields problems in understanding objectivity. Action at a distance, for example, poses the problem of physical matter reaching out and pulling itself as if energy goes and comes without any circular reference as found in pictorial examples using the magnetic field. Gravitational examples are pictured similar to radiation obeying the Inverse Square Law. Radiation does not return and how this concept applies to gravitation seems to be a contradiction with objectivity. On the other hand, the problem with curved space manifests its problems with assumptions about observation. Because light is a form of matter, modern physics says that light curves, but to call this curved space does not explain gravitational action on light. Light also curves around corners and no one suggests that the same principle is in play unless you are equating gravitation to graduated magnetic density. When a pitcher throws a curve ball, we see that the object curves in space and when we see sunlight curve against the curvature of the sun or even upon entering a prism, we see the curvature of light, but to say that space also curves, as if subjectivity and observation are objective, distorts reality and space curvature is not really explained.

Even though science cannot demonstrate a rational cause for gravitational curvature, we do understand that mass takes up space, that matter has form in space, and that matter moves within space. This is because we can understand the concepts of space and time in terms of velocity. With momentum, density and weight and other mass related concepts, we attribute a gravitational constant as the behavior of mass. It is this constant that evades understanding whether in Newtonian or Einstein physics. Just because we can put a constant in an equation does not mean we understand the cause of that constant. Just because something goes up and comes down and just because something curves does not answer the question "Why?" Even if we had a particle of radiation that modern physics attempts to define as a graviton, it does not give us the understanding equal to what we have as to space and time. Mass is illusive. We can conceptualize space and we can intuitively accept time, yet we come to assume an understanding of mass in conjunction with a non-understood mathematical constant. Even though we visualize solid-like marbles swishing in a bag, we do not understand gravity. It is only a matter of faith due to your observation. Gravity is a perfect example of faith working in conjunction with reason. Faith is not to have a perfect knowledge, for if we did we would understand. Observation teaches us to have faith, but science places faith in the subjective as some form of religious behavior rather than in the predicative as it is manifest in physics. Faith is predicated upon observation which is also a predicate and thus we have faith in gravity. We do not have knowledge of gravity, for knowledge understands the workings. This thesis attempts to provide that knowledge, while modern physics presents magic, strangeness and mysticism which are none other than a representation of a physical religion. Give me faith or give me understanding, but don't call relativity physics the ultimate truth. Believers simply chant voodoo that makes little sense to individuals outside their religion.

Just because we can pictorially imagine electrons, protons and neutrons does not give us understanding as to the cause of their attraction and repulsion. We only imagine them as space/time images. We can understand that an electron is smaller than a proton, but we cannot envision the concept of charge so we give it an attractive or repulsive constant based upon observation in the same way we define gravity. Just because electrons curve one way and protons the opposite, we do not understand this situation without the introduction of a magnetic field, but still we do not understand the why. The reason the magnetic field has a unique presentation is that it is circular. Both the electric field and the magnetic field have this ability to circulate each other, thus giving us a graphic understanding of mass as a reasonable compilation of the two fields. We still have the problem with gravitational action at a distance, but if gravity is an electromagnetic relationship we have duplicated the action of micro infinity to the action within macro infinity. We seem to accept this action on a microscopic scale but to say that a sun has a magnetic pull at infinite macrocosmic distances is unfathomable. Until we understand that the pull is local and not infinite magnetic properties can give us some understanding. Without applying the same local relationship to gravity we fall prey to the magic of action at a distance. This tradition has taught us to believe and not understand. It behooves us to unify the electro-magnetic relationship with gravity. Einstein attempted this direction, but modern physics preferred a magical paradox to the solution of understanding. Writing an equation to predict behavior does not satisfy our understanding. We need to answer why? We need to understand this gravitational constant inherent in density, momentum and weight.

By extending the equation of mass out to its individual space and time proportional measurements, the result suggests that all is space/time:

$$\text{MASS} = \frac{\text{GRAVITATIONAL CONSTANT}}{\text{ACCELERATION}}$$

$$\text{MASS} = \frac{\text{ACCELERATION} \times \text{Length / Height}}{\text{ACCELERATION}}$$

$$\text{MASS} = \frac{(\text{Length / Time square}) \times (\text{Length / Height})}{\text{Length / Time square}}$$

This may be simplistic, but it does illustrate a point. As indicated in a previous chapter, space and time must have a zero point of measurement. In the case of mass the zero point treats the total mass as if it was at this zero point or what physics refers to as the center of mass. The zero point of both velocity and acceleration are at right angles to the zero point or center of mass. This is indicated by the use of height and is also the essence of two mass points in respect to each other such as the earth and moon having a space difference in terms of height rather than in terms of linear or angular motion. All of mathematics is essentially points in space and motion in time. The right angle relationship gives us the gravitational constant and is considered a technological feat in predicting the actions of a ghostly force. It does not, however, give us understanding as to the cause of this behavior because mathematics is only a compilation of space/time units. Observation in conjunction with a physical time piece provides a gravitational constant, but that constant is simply a space/time relationship. It is like predicting something through faith without the past and future understanding of how. This is the nature of mathematics. It predicts, but does not explain. Mathematics is a predicate reality just like observation, measurement and faith. Mathematics is not knowledge until we understand mass in the equation in the same way we understand space and time. Many still struggle in the understanding of the space/time proportion. Until we fully understand mass in behavior terms, mathematics can only predict. Just because we know the *'what'* in a prediction does not mean we understand the *how* of things. We learn the *'what'* in life by observation and experience, both of which are predicate realities. The *'how'* is subjective or the rational process of intelligence. The irrational is simply concocted belief systems based upon conjecture and motivated by the denial of responsibility to reason. In more advanced subjective terms the *'why'* becomes that vision into cause in the same way *what* becomes the vision into the effect. The more we understand the *why* by keeping eternal concepts in perspective, the more we can claim knowledge rather than faith about the objective world. Modern science avoids eternal concepts in the same way they avoid God. Science simply gets caught into a web of mysticism in the same way a religion of a false God gets caught into a web of magic.

    Just as a better understanding of *how* and *what* help us understand the objective world, so also is it possible to better understand the space/time relation in order to derive our meaning of mass. The last equation breakdown illustrates this. It is also why we have not as yet come to grips with the geometry of gravitation. You might say that we have to some degree come to an understanding of inertial mass as action and reaction due to the quantitative nature of atomic units, but the gravitational constant affecting these units has no geometrical cause for intelligence to lay hold of the *how* in geometrical terms. We simply predict out of experience. Again we come to magnetic properties as the only force that we can geometrically understand. The gravitational force of action at a distance does not have that blessing. Physics needs to look into the geometry of field theory and not just accept the ability to predict something. Curved space will not due, because it has no cause. To say that gravity curves space is a circular argument. We are simply back to the beginning when we ask, "What is gravity." The attempt of this thesis is to give understanding and still maintain both Newtonian and Einstein gravitational predictions.

Though intelligence gives mass a generalization of spatial volume, it must be understood that, one volume of matter has an apparent density different from another equal volume. Volume and density, in other words, are inversely proportional. If the mass units remain constant then the volume decreases proportionally to the increase in density or the density decreases proportionally to the increase in volume. Density, in other words, evolves from a pure conceptual proportion as related to a measured gravitational constant.

The apparent inertial concepts evolve from the gravitational experience. By using a balance scale, for example, a mass on the moon weighs the same as it does on the earth. We understand this from the logic of the equality of mass probably being the same regardless of its change of location. You can change the constant and it affects all conditions equally. We speculate that this is due to the inertial or objective properties. Inertial mass is not relative when considered to be some sort of atomic units. By using a spring scale, however, mass on the moon weights less than the same mass on earth. This is the action of a differing gravitational constant between the moon and the earth. Gravitational mass is relative to each planetary object or a change in the constant. Since a balanced scale measures the same mass on the moon as equal to the same mass on earth suggests non-relative or inertial mass. This is due, not because the mass is not working against the gravitational constant, but against a like particle. This is the objective nature of mass. It is not relative because it is inertial.

To reach beyond mathematical mass points and spatial derivatives, a better understanding of the measured relationship between the variable gravitational mass and the invariant inertial mass is eminent. As suggested, a reliable field theory using the magnetic field as the foundation may some day reveal both the correlation and the differences?

Einstein worked on a unified field theory in his later years, but unsuccessfully left it to others that followed. A good field theory not only could explain the force of gravitation and all other action-at-a-distance forces in nature, but it might well explain why inertia gravitates or curves. Once all forces are better understood in controlled experiments, the measured inertia or resistance to motion will imply more emphatically that mass is ultimately objective. But now, the curvature of mass remains a mystery to most. We believe in curved space or the dinosaur philosophy of action at a distance. Even though we have not come to a better understanding of relative gravitational mass, our understanding of inertial mass is somewhat clear at this point. We have reached this point because of the conservation of mass.

Conservation of Mass

Unnoticed to most, history has taken the law of conservation in conjunction with a properly defined objective reality (mass being a divisible substance), and has used clearly defined terms and conservation in measurement to help search for the smallest unit of object existence.

The beginning of the search for the smallest unit of matter began with Democritus. Democritus, a philosopher about 430 B.C. called these yet undiscovered units *atomos*, from the Greek word meaning indivisible. From this the word atom finds its origin. Alone in his views, it was only the critical remarks in the works of his contemporaries that preserved the view of Democritus.

The Einstein Illusion

The atomistic philosophy of Democritus must have affected the alchemists who believed all matter to be made up of a single substance. They thought this substance became the four elements - earth, air, fire, and water. No doubt the philosopher chemist of the time realized that with the aid of fire the simplicity of transforming water to air came to light. Robert Boyle (1627-1691) found that air could be compressed or expanded. Something common resided in nature's substance, but the complexity of too many elements arose in the theory of matter. Reason implied that there must be a smaller unit.

The law of conservation of mass, maintained by Lavoisier, and a desire for order in existence caused Joseph Louis Proust (1754-1826) to always find the elements in fixed proportions; hence, the axiom of proportion in conjunction with sound conservation of measurement reduced mass to one single concept called the atom. He found that in copper carbonate, 4 atoms of oxygen to 1 gram of carbon appeared with every 5 grams of copper. One might conceptually conclude that for every atom of Carbon, 4 atoms of oxygen stood by or that an oxygen atom weighed 4 times as that of a Carbon atom. As shown below, neither explained the situation; but the axiom or law of fixed proportions pointed Proust in the right direction. Using Carbon Carbonate, the following comparison resulted:

Carbon Carbonate Fixed Proportion	= Carbon = 12 grams	+ Oxygen + 48 grams	+ Copper + 60 grams
Atomic Weight	12.011	15.999 15.999 15.999	63.54

Total Atomic Weight	= 12.011 + 47.997 + 63.54
Total of 5 Atoms	= 1 Carbon + 3 Oxygen + 1 Copper

The ultimate result turned out to be 3 oxygen compared to 1 carbon and a very heavy copper atom to the other 4. But the relationship of the fixed proportions to the total atomic weight of the now 5 atoms reveal that Proust's law closely approximated the truth for such methods of measurement. Based upon this law of fixed proportions, John Dalton (1766-1844) advanced the modern atomic theory. Dalton believed in Democritus and the indivisible atom. The law of fixed proportions by measured weight, however, did not give the actual mass of each atom; but it allowed Dalton to assign each element an atomic number proportional to each of the other elements. Being the least in mass, Dalton gave hydrogen the value of 1. Since hydrogen and oxygen formed water in the proportion of 1 to 8 (by measured weight), and since hydrogen received the value of 1, then oxygen should equal 8. But when the physicists split water, they found that for every liter of oxygen 2 liters of hydrogen remained. The proportion by measured weight differed from the proportion by measured volume. To satisfy both proportions it became necessary to give oxygen the value of 16.

Water	=	Hydrogen	+	Oxygen
by weight		1	to	8
by volume		2	to	1
Atomic weight		1	to	16
Water molecule =		2 atoms Hydrogen +		1 atom Oxygen

The law of conservation has given every element a determined weight by a volume proportion; but the kinetic theory of matter in conjunction with the study of gases led to the actual size of the atom, space within included. According to the theory, a gas accumulates as a great number of particles (or molecules) move and collide from all directions. The average speed of the kinetic motion of the molecules determines the temperature of the gas. To any definite temperature a corresponding definite average kinetic energy per molecule remains. Take two vessels containing equal volumes of different gasses. Both gases have the same volume and temperature if both identical vessels have the same identical pressure. If the volume, temperature, and pressure are measured equally for different gases, the number of molecules moving in each must be the same.

Sometimes a gas will exist as a single atom, but Amedeo Avogadro (1776-1856) carefully used the word molecules.[2] Avogadro suggested that all gases at a given temperature and pressure contained the same number of particles per unit volume. The particles may or may not be atoms, but the number of molecules in a gram of any gas could be called Avogadro's number. Avogadro's hypothesis expressed that equal volumes of all gases contained equal numbers of molecules under equal conditions. From the equations derived by James Clerk Maxwell (1831-1879) and Ludwig Boltzmann (1844-1906)

> ...it was possible, by making some reasonable suppositions, to calculate what Avogadro's number might be. This was done by a German chemist, J. Loschmidt, and turned out to be approximately six hundred billion trillion—a large number, indeed.[3]

Atoms then turned out to be the same size but not the same weight. If a hydrogen atom had more space or emptiness within its boundary than the oxygen atom, it implied that the atom could be divided and Dalton prematurely gave the smallest unit the name *atom*. The divisible atom eventually found its place as a composition of electrons, protons and neutrons. Dalton's eagerness to name the basic unit prevented his acquiescence of greater simplicity.

Some elements discovered had almost the same chemical properties, but differed in physical makeup. Neutrons did not affect the chemical nature, but did affect the mass of each atom. These elements, called isotopes, possessed only one difference—that being one or more extra neutrons. The atomic weight merely indicates the average of the masses of more than one isotope of a given element. Assuming hydrogen to = 1 proton only, then for every 125 atoms of hydrogen, one atom with an added neutron would yield an average atomic weight of 1.008 for hydrogen. Until the proton number changed the element remained chemically the same. With this in mind the meaning of the atomic number differed from the atomic weight. The atomic number equaled the number of protons while the atomic weight approximated the average number of protons and neutrons per atom. Although every isotope with extra neutrons existed in very small amounts, the ones that did have extra neutrons affected the average atomic weight in chemical measurements.

Objectivity in Matter

Element	=	Neutron Weight	+	Proton Weight	=	Total Isotope Weight	Average
Hydrogen		0	+	1	=	1	
Hydrogen		1	+	1	=	2	1.008
Helium		1	+	2	=	3	
Helium		2	+	2	=	4	4.002
"							
"							
Carbon		6	+	6	=	12	
Carbon		7	+	6	=	13	12.011
Nitrogen		7	+	7	=	14	
Nitrogen		8	+	7	=	15	14.006
Oxygen		8	+	8	=	16	
Oxygen		9	+	8	=	17	
Oxygen		10	+	8	=	18	15.999
"							

Because of the extra neutron in an occasional carbon-13, the average atomic weight is 12.011. The same explanation applies to nitrogen. Oxygen and some of the heavier elements with few isotopes do not fit a perfect pattern. Assuming oxygen to equal 16, the average atomic weight should be something like 16.004 with oxygen 17 & 18 pulling up the average. Due to a disagreement between physicists and chemists in determining the atomic weights, in 1961 they agreed to determine atomic weight on the basis of allowing the carbon-12 isotope to equal 12.00000. This tied all the atomic weights to a new number. As shown above, the atomic weight of oxygen came under the new system as 15.999. But still 15.999 did not equal 16. If the lowest oxygen isotope equals 16, it is proportionally impossible for oxygen to have an average atomic weight of less that 16.000. Regardless of the system of weight assignment, the average atomic weight of any element should come out greater than the lowest isotope.

The answer came with the suggestion that a proton weighs less in iron than in the lighter element oxygen. Close examination of each element clearly shows a gradual loss of weight with each proton and neutron as the element progresses up the atomic scale to iron. Continuing up the scale produces a gradual increase in weight per proton or neutron. The elements can be arranged in such a way that iron resides at the bottom of a curve. The elements from both ends of the atomic scale roll down hill, so to speak, from each end giving up mass until they all meet at the bottom to reside as the most abundant element iron at the center of the element chart.

A mass conservationist might argue that **Figure 7-1** explains the fluctuation of outer particles such as the electrons, but in studying the electron, the disappearance of mass demonstrates that electrons also loose weight. There had to be an accounting of the missing mass.

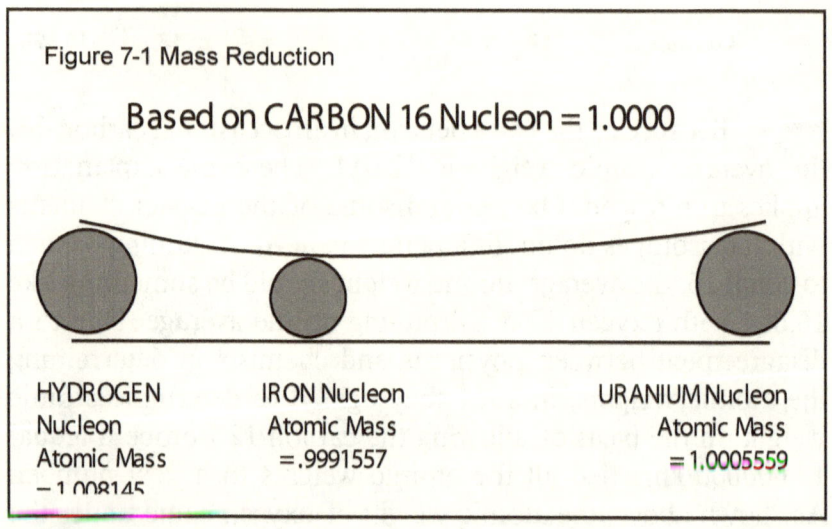

Figure 7-1 Mass Reduction

Based on CARBON 16 Nucleon = 1.0000

HYDROGEN Nucleon Atomic Mass = 1.008145

IRON Nucleon Atomic Mass = .9991557

URANIUM Nucleon Atomic Mass = 1.0005559

Mass-Energy

In Einstein's second paper, he related mass to energy and gave us $E=mc^2$. If it can be understood that gross mass is the geometrical pattern of smaller orbiting mass particles creating the angular motion of a closed sphere, then $E=mc^2$ says nothing strange or complex. But if the relativist equates mass and energy, just as space and time are equated, the new accountants of objective truth have failed to picture a geometrical model of reality where mass and energy are inversely proportional rather than convertibly proportional. Just because the equation process can be inverted does not mean that inversion means conversion.

Objectivity in Matter | 138

Mass-energy inversion does not imply substance at one moment and non-substance at another. Energy and mass coexist, but differ as to reality. If the relativist is determined to maintain a conversion and not a relational change in direction of two separate realities, so let them believe in magic. The conversion is far better manifested as the conversion from angular energy to linear energy rather than one reality to another. It can also be said that the accompanying mass can be termed angular moving mass as being converted to linear moving mass. In this respect both mass and energy exist together without trying to convert one to the other. We conclude that light is not energy only but rather a piece of mass spiraling from a stationary orbit into a fast moving linear photon. Thus there is a loss of both energy and mass from an electron and other particle when light leaves. Conversion is a ridiculous concept. Mass and energy change is but from one mass motion to another motion of the same substance—that substance being the smallest unit of mass or the smallest unit of light. It does not matter which. Rest mass defines best the angular motion, while light defines the linear movement. Both are the same inertial substance, or as this thesis will maintain, the same inertial photo-electric particle constantly maintaining a field relationship to the magnetic field. A whirl wind of orbiting photo-electric particles measures a single ponderable mass particle. A loss in mass comes when a portion of angular moving electric particles (fields) spiral off as linear moving photons around a magnetic field at rest. Simply put, a portion of the total angular momentum of the system is converted to linear momentum. This is not a conversion of one reality to another. It is a conversion of angular momentum to linear momentum and also a conservation of both angular mass to linear mass. Likewise it is a conversion from angular (total) energy to linear (kinetic) energy. Half of the energy and momentum recoils with the particle such as an electron and the other half of the energy and momentum leaves with the photon as kinetic energy. The mass is not equally distributed between electron and photon but the total mass is conserved. To say that light has zero mass is not looking at the equation properly. We mistakenly think that the electron is a stationery solid particle and the photon as some sort of pure energy without mass.

The mass particles of light, lost from the atom as it moves up the atomic scale to iron, returns as an absorbed photon as the atom continues to move up the other end of the atomic scale to the very heavy elements. The coming and going of light particles equal the missing atomic weight of the individual atoms, for the space within every atom holds many electric particle fields that are held together by surrounding magnetic density fields. Each maintains a certain particle density and strength. Each atom exists as nothing more than a whirlwind of photo-electric particles creating several complex electric and magnetic field arrangements around a net magnetic axis. This concept of particle light provides a generalized step in understanding the objectivity of matter and thus its ultimate indivisibility without disappearing and appearing magically.

The relativity mass was developed out of atom splitting experiments. The same experiments have done well to demonstrate the non-relativity of mass. This was most clearly emphasized when the atom blew apart at the seams. Experiment had demonstrated that the Democritus atom proved to possess various numbers of electrons, each having a certain energy level calculated to be at very large distances from the atomic nucleus itself.

> The fact that alpha particles could pass through 20,000 gold atoms as though they were not there was strongly in favor... of an empty atom—an atom that is made up of nothing more than a scattering of light particles.[4]

Also, when it was discovered that a negatively-charged zinc plate lost its charge if exposed to ultraviolet light, and that a positively-charged zinc plate remained unaffected, this created the realization that the negatively charged electrons were drawn out of the individual zinc atoms by the absorption of photons making the electron heavier and able to separate from the atom by perhaps creating its own magnetic field as it no longer needed to share with the atom. The electron gained mass by the addition of a photon bundle. Inversely, additional electrons can be forced

Objectivity in Matter | 140

onto atoms by the removal of light particles thus perhaps limiting their own magnetic field and they fall to a more dense magnetic space closer to the atom until they find equilibrium—the more photons they lose the closer they get. The number of electrons does not affect the characteristics or chemistry of the element, but the charge is affected with a change in number. With the different mass levels of the electron, one could then speak of several electron orbits or shells. Depending upon the number and also the location of the electrons as they moved from one level to the other, a certain piece of light absorbed or emitted, provided the stability of the atom. This understanding of the electron and its shell position completed the periodic table.

Keep looking but never assume the end until but one particle becomes the smallest unit of objective reality. Until unity arises, everything is divisible. With the atom coming apart with three basic units the question arises, how much emptiness lies between each sub atomic particle? Does space exist of itself or is there no space of which there is not some form of matter? Far too many questions with only one simple assumption: the smallest particle of light may just as well be the smallest particle of objective matter. Photons can also be considered quantum bundles of the electron's photo-electric particle field. The atom seems solid, but rather mostly empty space resides therein. If objective space is preferred, call it "magnetic space." Together, the atom composes the harmony of a complex interaction of electric fields and magnetic space or better yet, another form of objectivity called a magnetic particle field.

In returning to the classical square of reason, picture a sealed test tube of gas molecules with each molecule possessing their own angular momentum. If for some unexplained reason a very small portion of the molecule brakes down through the conversion of angular momentum into small enough packets of linear particles that permeate the glass test tube, we have an apparent disappearance of mass. But if the smaller packets passing through the glass exit as bundles of spiraling particles moving in a linear spiraling path, then these photon bundles escape as the smallest aggregate of matter. If these bundles segment the electric field, they can appropriately be called bundles of photo-electric particles. A certain number would make up a given quantum or photon size. **Figure 7-2** will illustrate that the greater the number of photo-electric particles per photon bundle, the greater the apparent frequency created by the total angular velocity and the greater the total mass of the photon spiral. In place of every word "photon" or "photon bundle" insert the words "photon spiral" and classical thinking is reborn. Just as space lies within the prematurely named atom, the empty structure of magnetic space also characterizes each nuclear particle. Deeper still, space permeates the photon. We need to stop thinking of solid particles and mathematical centers of mass as the end. We need to stop relying on the equation and construct models that make sense and still agree with mathematics.

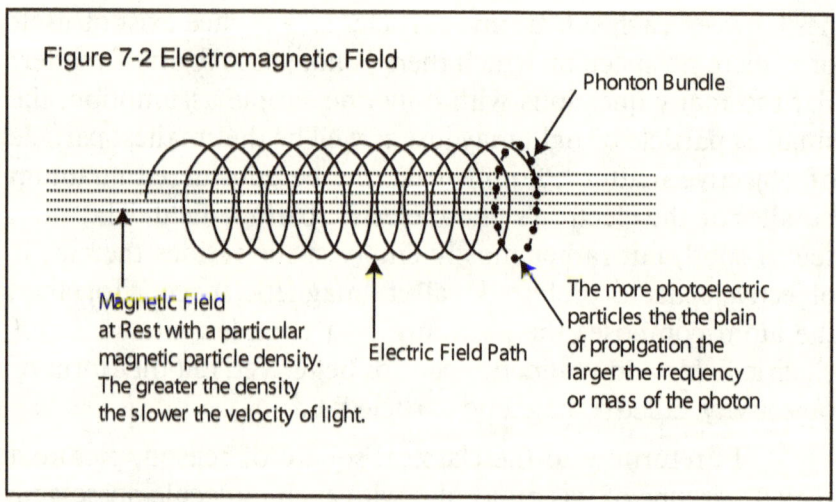

Figure 7-2 Electromagnetic Field

Phonton Bundle

Magnetic Field at Rest with a particular magnetic particle density. The greater the density the slower the velocity of light.

Electric Field Path

The more photoelectric particles the the plain of propigation the larger the frequency or mass of the photon

Relativity says that light has mass only at rest. Rest energy within the atom puffs into what the relativist calls "pure energy of motion." This pure energy accelerates from zero to the velocity of light. During that acceleration, light increases in mass as some might say until its mass equals the original mass within the atom. Change the word "pure" to "linear" and the word "rest" to "angular" and a better meaning develops without changing $E=mc^2$. That is the essence of a good model.

Light, traditionally measured as a wave of pure energy, mistook the particle of light in motion as nothingness in motion. The wave function of light does not measure the up and down or back and forth oscillations of a medium, because light does not travel in a medium. The geometrical wave concept is simply the electric field or photo-electric configuration. On the other hand, the mathematical or experimental wave function of light is nothing more than the photon's refraction index within a change in magnetic field density. The denser the magnetic field the photo-electric spiral enters or leaves, the greater the refraction. This also reveals the apparent change in direction.

If the magnetic field density on both sides of the traveling photon remains equal, the increased density ahead decreases the velocity of the photon. But if the magnetic field graduates from left to right yielding a greater density on one side of the path of travel, the photon will bend into the denser portion from the apparent slowing of the photo-electric particle's reaction to the one side. Light bends as it moves from one magnetic medium to another, not because of any wave movement. The spiral nature is really the only wave effect, but it is the magnetic density that really causes the change in direction.

Nothing ever discovered is absolutely new. The model of light, here described, may very well be in the pages of some theorist's writings. Faraday certainly postulated the beginning of this theory. No other theory solves the quantum wave-particle duality and the relativity paradoxes. There is not as much as a single explanation in popular or even text-book literature equal to this solution.

Field theory has arisen only as a mathematical explanation of lines of force. But if the field is a particle reality, then a very strange thing will happen to all of relativity including the strangeness of action-at-a-distance and curved space gravitation. Simply put, action at a distance is not a cause, only a mathematical convention. The earth's curvature or gravitational pull is a local phenomenon upon each mass of electric fields as they react to the vertical magnetic graduation.

When energy travels from electric particle to particle, for just a moment in time, mass/energy will reside in the magnetic field between. For this reason, take your measurement fast enough and you will notice that there seems to be a momentary loss of mass/energy between the two fields. The photo-electric particle field is only taking a quantum leap between particles. There is no space in which there is no magnetic properties and the greater one finds electric properties the greater will be the density of magnetic space.

Objectivity in Matter | **144**

The problem with objectivity is in the act of observation over time. Once you necessitate the use of observation you need to have a logical basis in the mind that observation can show agreement. One does not conclude in the mind by what one sees. It is the other way around. It is the objective world that receives the action of observation. You cannot disavow logic just because observation appears to distort.

Samuel Dael

(Endnotes)

8. The Predicate Nature of Energy

> There is an intuitive feeling that one will not be able to get something for nothing. It therefore seems proper and orderly to suppose that the universe possesses a fixed amount of something or other (such as momentum or energy) and that while this may be distributed among the different bodies of the universe in various ways, the total amount may neither be increased nor decreased. (Italics added)
>
> -Isaac Asimov[1]

The Mechanical View

Nicolaus Copernicus (1473-1543) first suggested that the earth rotated upon its axis and moved about the sun. Those who objected to such doctrine felt it impossible for the earth to move; for if one jumps into the air the earth would most certainly move beneath and anyone jumping as such would land several yards distant. Those who supported Copernicus felt that on the other hand, a leaping man would simply move with the earth having the same conservation of momentum and therefore would come down at the same spot he jumped from.

> Galileo pointed out that an object dropped from the top of the mast of a moving ship fell to a point at the base of the mast. The ship did not move out from under the falling object and cause it to fall into the sea.[2]

The reasoning behind the conservation of momentum stems from Newton's laws of motion. The mass of the moving body and the apparent energy it carries are conserved. If the mass of a moving body suddenly disappears without direction or if the motion accelerates without cause, the universe would be in total chaos. We thus have the laws of conservation of momentum and its derivative, the conservation of energy.

The idealistic pool table gives a good analogy of conservation in momentum. As the billiard balls move across the surface of the table, each inertial force will manifest its own direction. An experienced pool player instinctively anticipates two forces. One inertial force maintains a horizontal direction and another gravitational force creates the vertical force towards the earth. This gravitational energy passes to the ball and keeps it in contact with the table surface. The kinetic energy that each ball possesses comes from the direct impact of the other billiards or the original thrust of a cue stick. The air and surface friction in the moving path of the billiard also should not be forgotten. All forces add to the total momentum and kinetic energy of this system.

Consider only one force, the force of impact of one moving billiard and the equal and opposite reaction of another. This fulfills Newton's third law of motion. The inertia possessed by each billiard also maintains a constant velocity or state of rest until acted upon. This state of inertia defines Newton's first law. Finely, Newton's second law explains any form of acceleration on the pool table. This law states:

> The acceleration produced by particular force acting on a body is directly proportional to the magnitude of the force and inversely proportional to the mass of the body.[3]

The equation can be written as:

FORCE = MASS x ACCELERATION

Or you can invert the equation as:

ACCELERATION = FORCE / MASS

This mathematical proportion states that the greater the force or the less the mass the greater the acceleration. In other words, if no net force acts on a body, it undergoes no acceleration and must therefore either be at rest or traveling at a constant velocity.

It follows, then that the second law of motion includes the first law as a special case. If the second law is stated and accepted, there is no need for the first law. The value of the first law is largely psychological. [4]

The experienced mind will visualize that a stationary billiard relative to the table has zero momentum and rest energy, but a moving billiard has positive momentum and kinetic energy. The quantity of momentum mathematically differs from the quantity of energy by simply squaring velocity. Both momentum and energy pass from one billiard to the other at the moment of impact. Both momentum and energy share the same center of mass that moves in a constant motion. It is the individual mass billiards that produces the oscillation or wave motion. Energy does not come in waves. It is constant, less the relatively constant friction for each reaction to the last billiard impact against the table edge. Energy does not wave. It is the billiards that do all of the waving or in the above case—oscillation. In both sound and the motion of billiards a wave motion is not exactly what is happening. The wave is only apparent because the action of the mass units yields the wave function. In a real sense the wave pertains to the medium, but energy is a constant that pertains only to the net predicate action of all the mass involved.

Figure 8-1 represents a graphic example of the transfer of energy and momentum as an apparent constant motion. Energy and momentum in this sense are not objective because the center of momentum or energy does not stay with the geometrical center of a single objective billiard. The center of momentum is continual whereas each objective mass oscillates.

The fact that no single particle of mass traverses the entire distance of a chain of colliding billiards demonstrates that energy and momentum are not objective. If you have a series of billiard balls at equal distance from each other, and if you strike the first, the second will receive the apparent energy and momentum of the previous ball. The first ball stops and the second will take all motion to the next. Once the second passes the energy to the third, it stops and the third takes up the motion. The agitated motion transfers from billiard to billiard

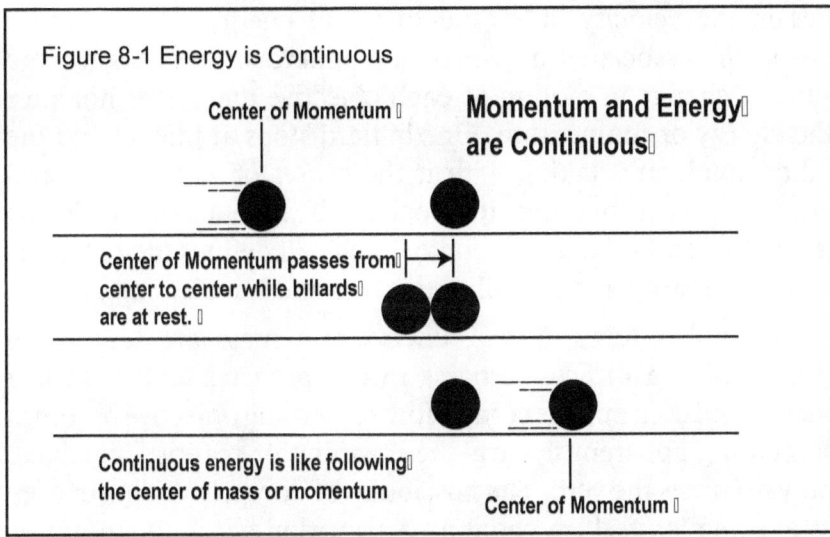

Figure 8-1 Energy is Continuous

until the last billiard is left with all the kinetic energy less the total loss of friction. This energy manifests a continuous moving center passing through each geometrical center of mass. Under the collision of perfect spheres, this center of momentum passes continuously from the physical center of one billiard to the point of contact and finely to the physical center of the second billiard. This concept demonstrates that the local wave motion is described by the change in motion of each billiard, but the energy and momentum are continuous. The same applies to a water wave. The wave motion towards the shore is a motion of appearance and not actual water molecules. By examining an individual water molecule, it could be shown that each basically moves up and down and not necessarily toward the shore. Even the last section moves not so much up and down but back and forth upon the sandy beach. The change is due to the energy hitting the beach. In other words, the wave of mass is local while the energy is continuous until the inherent energy hits the sandy beach causing an opposite reaction. Like a spirit without substance, energy is motion, but this spirit cannot exist without a mass to carry it. Energy, therefore, must be apparent and predicative. The amount of energy passing from billiard to billiard is proportional to the unit sizes of the billiards. Since mass comes in units, so does energy. The continuousness of direction

The Einstein Illusion

lies in the velocity of momentum and energy. The quantum lies in the associated mass. As one billiard strikes another, the sudden change in motion of each objective mass does not alter the energy or momentum. One billiard stops at impact and the other accelerates taking with it the center of momentum and energy. Energy signifies apparent predicative action while the individual stop-and-go motion of each ball defines the nature of objective reality and its ability to carry and transfer energy.

In Einstein's $E=mc^2$, energy and mass are said to be convertible—a difficult process to comprehend under the just mentioned concept of energy. How something objective becomes something apparent thwarts the meaning of reality. The object now becomes the verb. The possibility of comprehending energy as existing by itself, without an association with a quantum of mass, defies reason. Energy can easily be understood as an apparent motion of the objects involved, but to treat energy as objective in some sense, and not as the apparent motion of a mass point moving from particle to particle, boggles the mind. Yet, if we think of light not as pure objective energy but as a particle spiral that carries a quantum of mass with it, we ask, "How did the particle obtain such a velocity from a point of rest?" To accelerate from rest without cause, mimics the belief that the appearance of both momentum and energy come from nowhere. To solve the dilemma relativity says that a portion of pure energy of action comes from a portion of loss in objective-mass. This epistemology comes from the mind of religious concoctions by slightly twisting the meaning that the equation offers.

Why so many scientists accept this conversion from objective reality to pure predicate motion seams hurriedly academic to the spirit of reason and the spirit of conservation. Nuclear reactions and the reality of global annihilation say absolutely nothing about mass energy conversion. The equation $E=mc^2$ ought to speak only of a conversion from one energy of an angular direction to another linear direction just like the conversion of angular momentum to linear momentum.

The relativist says that the photon has a rest mass within the mass of the electron and that in order to obtain such high velocity the mass disappears into pure energy of motion. The rationalist, on the other hand, says, "A photon spiral never rests." Within the electron, the momentum is angular but outside it has linear properties because the magnetic field is at rest. The conversion becomes a change in direction not a change in reality. Light is mass and conversion is not necessary. General relativity, or the bending of light in a gravitational field, seams to explain the mass-like characteristics of light. Also, under the Photoelectric Effect, discussed in Einstein's first and second paper, light seems to have particle characteristics. But the quantum relativist emphasizes light to be pure energy without a piece of mass. This is not due to the equations of relativity, but comes from a change in reality or a magical change of terms. The only model that can be created to fit both the classical square of reason and all of relativity equations is the spiral helix model of photo-electric particles that bend upon incidence with a change in magnetic field density. This also satisfies quantum theory.

Field Refraction

When energy moves from one mass to another by direct contact as in action and reaction, the back and forth or up and down acceleration of mass particles follows the rhythm of a continuous moving energy. How should we classify indirect transfer of energy due to a magnetic field? Even weak gravitational and strong nuclear field forces engender energy transfer. Mass particles, large and small, set up magnetic fields. The distance between particles and their mass determines the potential energy passed through the fields involved. The sun sets up a field in which the earth resides. The angular momentum of the rotating earth, orbiting the sun, includes a magnetic field as part of the system. When considering light to be particle in nature, Newton's calculations would yield a similar curve or refraction as that determined by a comet glancing through the sun's field. For a moment the particle of light will give up to the field some of its momentum to the new system through a slowing of velocity. Einstein predicted this curvature of light at the velocity c, yet relativity calculations return a curved for light as being twice that calculated by Newtonian theory.[5]

Many experiments have been made in determining the degree of light curvature or refraction in a gravitational field. All of them more or less confirm the Einstein relativistic equation. Einstein's equation, written symbolically is:[6]

CURVATURE OF SPACE-TIME GEOMETRY	=	GRAVITY	x	MASS DENSITY of MATTER IN SPACE-TIME

General Relativity defines the curvature of space as the product of Newton's gravitational constant G and the net mass density of the system. General Relativity theory requires that mass, planet, or comet fall freely as straight as they can go through curved space—the greater the mass density of the system, the greater the curvature. The mass density is not in the individual components, but in the system as a whole.

Newton's calculations assume instantaneous action at a distance through empty space. The axiom states, "Matter attracts matter." General Relativity uses different words: "matter causes space curvature." Einstein's equation was not the first to modify Newton's gravity. Maxwell's equations of electric and magnetic effects questioned Newtonian gravitation. Perhaps not too great an oversimplification, the far right portion of Einstein's equation pointed to the real essence of Maxwell's space—the greater the mass density of matter, the greater the magnetic field density. No one considered that.

Field geometry and Maxwell's field equations actually do away with empty space. Instead of matter attracting matter, the external field causes the change in direction. A field theory describes a smaller particle's field as being redirected when passing within the confines of a larger particle's field. An uneven change in magnetic field density will yield a proportional change in direction. The basic reason why light bends twice that of gross matter will be explained. It will have more to do with geometry than relativity. Here, the point will be made that the reason for any curvature at all can be explained without the curvature of space/time. Curvature is due to the irregularity of magnetic space, which is a direct variable result of the density of ponderable matter.

Just as Einstein imagined as a young boy what it would be like to travel within a photon. Imagine you are inside the spiral photon mentioned earlier and you are glancing past the sun. Imagine extending your hands to a distance equal to the circular diameter of your photon. As you face the direction of travel, you see that the photo-electric particles (electric field) circles about you counter-clockwise from the right hand up and over to the left. Suppose that you can also measure the density of the magnetic field as one would measure the wind through your fingers. As the magnetic field at relative rest passes through your hands, you skillfully notice that the magnetic density to the left hand exhibits a stronger effect than the right. You attribute this to the sun, which sits to your left. The further to the right and away from the sun, the weaker the magnetic density will be. You could just as well assume that you are at a fixed point in space with a circulating electric field and that the magnetic field moves through you at the velocity of light. Since you are moving, understand that the denser time relationship of the magnetic field upon the left hand causes the curvature. Not only did you lose velocity when you entered the sun's magnetic field, but your left side lost a little more than the right. In order for the reaction time of the electric field on the denser left side to equal the less dense right side, the momentum of your photon spiral turns into the sun. Relativity would say, "Falling through curved space".

Without changing the calculations of Einstein's equations, change only the epistemology:

CURVATURE OF LIGHT AS TO VELOCITY	=	VARIABLE MAGNETIC FIELD DENSITY	x	MASS DENSITY OF MATTER AS TO VELOCITY

The use of space/time is not a relative concept. It is the velocity of the subject of study. In mathematical prose the equation reads: The greater the change in the variable magnetic field density perpendicular to the path of travel (gravity) and the greater the net, mass field density to the velocity, the greater the curvature of light. While the net field density appears greater as the field density increases relative to the velocity, the gravity or variable field density exhibits a greater uneven cross section in the path of light. The greater the change in density from side to side in this cross section the greater the force of gravity toward the dense side or the greater the curvature. The potency of the whole matter lies in the spiral nature of light and how it reacts to a perpendicular variation in magnetic space.

The constant velocity of light in a vacuum structures the whole substance of relativity. But as judged from the above, light indeed moves slower and curves near the Sun. "Either that or Venus jumps!"[7] Watch the planet Venus by radar as it passes on the far side of the Sun; it will lurch several miles farther from the Earth and then swerve back again. The radar astronomers see nothing more than that their pulses slow down as they pass near the Sun. This assumes that radar has a specific velocity relative the density of a medium. Whether that density is magnetic or gravitational, can anyone describe the difference? Similar results have been obtained with spacecraft. Signals sent from the Earth to Mariner spacecraft far off in the Solar System, responded with a slower signal as they returned through the sun's gravitational field. **Figure 8-2** illustrates the relationship between a light source and a relatively small light particle as it passes through a variable magnetic field (gravitational if you prefer).

Due to the increase in field density surrounding the small particle, the light particle moves slower. In conjunction with this increase in field density, the side-to-side cross section (gravity) creates the curvature. This point of view may not be acceptable, but based on field theory; warped space should recede in shame.

Relativity describes the curvature of light as follows:

A massive body squeezes and deforms space in its vicinity. Space is "denser" in the vicinity of a massive body and so

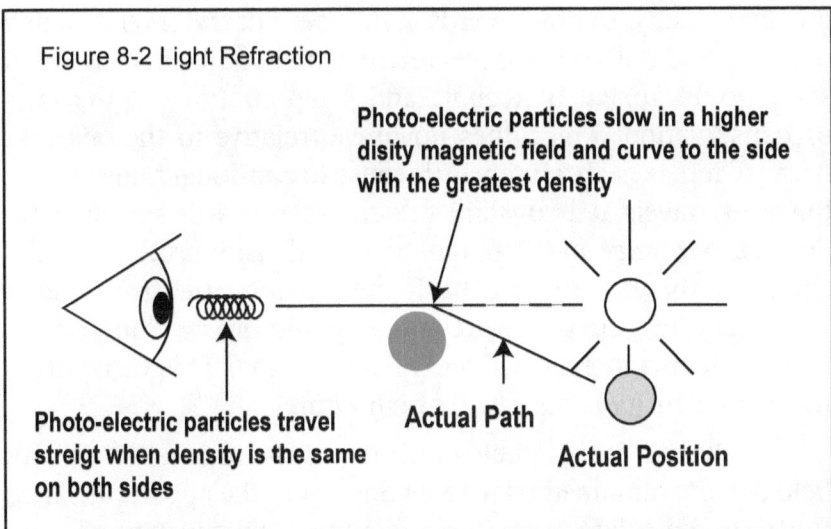

Figure 8-2 Light Refraction

Photo-electric particles slow in a higher disity magnetic field and curve to the side with the greatest density

Photo-electric particles travel streigt when density is the same on both sides

Actual Path

Actual Position

light seems to travel more slowly there, when seen from afar.[8]

The use of the words "squeezes and deforms space;" denote General Relativity representations of reality. Meaning could just as well be expressed more consistently with Euclidean and Faraday space. A far better description would be that *a massive body sets up a magnetic field in its vicinity. The field is denser in close proximity and so light travels more slowly there*. Density applies to the magnetic field, not space. Both statements have the same mathematics, but warped space desperately trades the physical existence of a field for the mystical aspects of warped nothingness.

The Special Theory of Relativity retorts that light has a constant velocity in a vacuum to all observers regardless of position or constant motion. This conclusion came by the direct result of the Michelson-Morley experiment. Now if light has a constant velocity in a vacuum here on earth, what of the velocity of light in a vacuum near the surface of the sun? It must move slower—General Relativity demands it. Now if light has a certain velocity in a vacuum on the earth's surface due to a weaker field density and a much slower velocity near the sun's surface due to a stronger field density, what about outer space far away from both earth and sun? The calculations should yield a greater velocity. In other words, light does not travel relative to a medium and it does not travel relative to its source because twin stars would appear to wobble and jump contrary to the laws of conservation. Light does not travel relative to the observer because it makes absolutely little sense to a rational mind. Light, however, travels at a constant velocity relative to a specific field density. A change in the magnetic field density produces a like change in the velocity of light. If the density varies from side to side, everything curves. If gravity obeys the Inverse Square Law, why not let magnetic density exhibit the same? This variation in magnetic density is the cause of curvature.

During the Michelson-Morley experiment, the earth's field density remained relatively constant to the apparatus. Thus Einstein's postulate prematurely assumed the constant velocity of light. Perform the same experiment on the surface of the sun or in far distant space and the Michelson-Morley experiment will still yield a negative result, not because the velocity of light moves at the same velocity from one position in space to another, but because the velocity of light remains constant to the apparatus if the net field density moves with the experiment. Conjectural error resulted when using the Michelson-Morley experiment to determine the constant velocity of light rather than simply debunk the ether theory. A sense of reason about a physical magnetic field would have removed every aether concept.

Einstein derived the constant velocity of light from Maxwell's equations and not the Michelson-Morley experiment. He only later referred to the experiment to more easily explain the Special Theory of Relativity. Maxwell's equations defined the result of experiments done on the surface of the earth. Naturally the field density here has determined that light will travel 300,000 meters per second in a vacuum. The same cannot be assumed to be the case in the far reaches of outer space. Atomic clocks, however, can move in magnetic space more easily than a laboratory. If light moves slower at the surface of the sun, so also will the atomic clock.

If a photon spirals as a bundle of minute photo-electric particles changing course upon entering a dense magnetic field at an incidence greater than ninety degrees. This is refraction proportional to the mass of the moving photon and the variable degree of magnetic space. All systems with external magnetic density alter the angular momentum of the photon. Space does not warp the motion of things. Bluntly, it is the magnetic field. Warped space becomes a denial of the laws of conservation and a rejection of the magnetic field as a property of particles.

Quantum Particles

Since the Michelson-Morley experiment rejects the idea that light cannot be a wave in a rigid medium and that light appears to have a constant velocity relative to the observer, Einstein's Special Theory of Relativity tries to explain. Space became warped in order to slow time sufficiently allowing the observer to conclude a constant velocity. Light became a wave of two oscillating fields with little need of a rigid medium. It should have been considered that the medium could be none other than the magnetic field that permeates space at various densities; and that the wave was not a wave of two oscillating fields, but an electric field that spirals around a magnetic field at rest?

Now if the electric field can be considered to be a field of spiraling particles of the smallest but equal bits of objective reality, a photon helix could be both a wave in that the frequency depends upon the magnetic field density and yet still be a particle in that the total mass depends upon the number of photo-electric quantum in the spiraling helix.

The debate as to whether light behaves as a wave of energy or as a particle can be resolved if light manifested the geometry thus considered. Instead, nothing more than the idea that light travels as pure energy exists among most authorities at this writing.[9] Only when light falls upon a photographic plate does it become mass. This leap of quantum physics boggles the mind, yet most every realist or objectivist will reject a mystical form of duality by asking for an epistemological account. Many subjectivists depart much farther by the denial of any duality at all. Under this form, the observed particle does not exist. Its creation comes from the mind or by the act of observation. Even the mathematician's equations do not represent anything real. "Just a game of numbers," retorts the modern day quantum physicist. One secretly cannot help but look upon this view as a philosophical extravagance born of despair, and a denial of the reality of entropy.

> On foundations we believe in the reality of mathematics, but of course when philosophers attack us with their paradoxes we rush to hide behind formalism and say, "Mathematics is just a combination of meaningless symbols,"...Finally we are left in peace to go back to our mathematics and do it as we have always done, with the feeling each mathematician has that he is working with something real.[10]

The common person will prefer the reality of particles and if no medium exists for a wave to propagate in then all things must remain as real particle fields. The often-called wave pattern becomes nothing more than the refraction pattern the particles create as they spiral from one magnetic field density to another.

The commonly referred pinhole or slit experiment still endeavors to explain the wave properties of light. The question is often asked, "If a particle, how can light bend around corners?" In fact, the patterns on the photographic plate turn out to be the same ring or stripe pattern explained by wave mechanics. For years the nature of light has been mapped out mathematically as a wave. But unlike other continuous phenomenon, light reacts in jumps or quanta, just as if waves moved in certain size packets of particles. Up until Max Planck's (1858-1947) discovery that light, like particles, came in indivisible units called "quanta," light was considered a continuous wave in the aether to eventually calling light energy that moves in nothingness.

Now if each particle carries so much energy as it moves through space and if particles only come in unit sizes, then energy

The Predicate Nature of Energy | 158

can only be transferred in unit size or quanta measurements. This can be explained in the slit experiment. In **Figure 8-3**, it can be assumed that light photon bundles will fall randomly at all areas of the opening. Neglecting the fact of scattering where some particles would strike the edge of the pin hole and bounce to the outer parameter of the photographic plate, the majority would pass directly through the center without changing direction, would strike the photographic plate dead center. But what will happen to a particle that passes off center? One cannot suppose that the typical pin hole or slit itself is empty space, even in a vacuum one must suppose that the magnetic field density differs from the center to the edge of the hole or slit. If a spiral of light enters the hole off center, it must be realized that as it passes through, the photon spiral will travel slower due to the ever so slight increase in magnetic field density. Remember, when uneven magnetic field density from one side of the hole to the other exists, quantum equations take hold. When considering that the field density next to a molecule is much greater thus creating a graduated magnetic density far greater than in normal

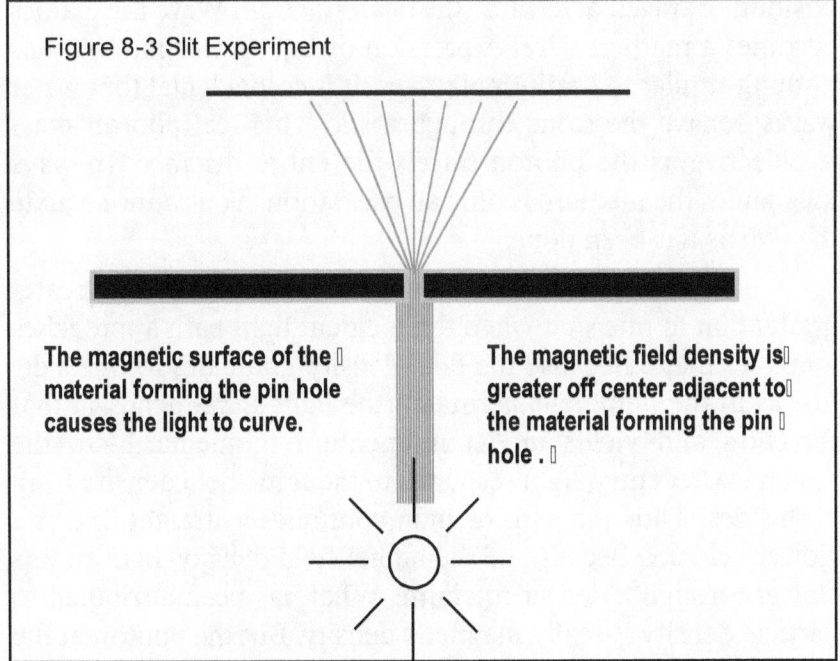

Figure 8-3 Slit Experiment

The magnetic surface of the material forming the pin hole causes the light to curve.

The magnetic field density is greater off center adjacent to the material forming the pin hole.

space, refraction becomes a matter of much greater curvature than normally expected.

Upon leaving the dense field area within the pinhole, each particle of light would regain its velocity consistent with the field density of the surrounding magnetically-filled space. And since the density will again measure equally on both sides of the path of travel, the photon will exhibit no change in direction—only velocity. The new direction determines at what point the photon passed through the hole and what the quantum size of the photon helix was. The slight blurring of the photographic plate indicates a combination of the two. The quantum size of each photon bundle could yield the various rings with the photons of the greater mass or closer to the edge of the slit refracting the greatest. The soft edges of each ring may be due to the random position of travel through the hole.

Now if photons came in an infinite number of sizes, the screen would be evenly distributed with exposed dots. But with only a finite number of unit sizes or jumps, a series of rings result. The field density, the mass of the photon and the position of travel, determine the net refraction. Wave mechanics becomes a mathematical expression of light refraction, but has nothing similar to traditional waves. It is coincidental that water waves behave the same through a slit. This real photon mass is objective as the photon travels the entire distance. In wave mechanics the medium is only an oscillation. You cannot equate the two as has been done.

Figure 8-4 illustrates how the field density has a greater graduation to one side when the incident light path approaches a greater mass. Because the field density graduates to one side, just as in the pinhole experiment, the light spiral bends in that direction and yields up its momentum momentarily to the system. After entry, light responds to the same field density from both sides. Thus the path of light continues a straight line at a slower velocity. Because of the higher field density, light travels slower through a denser medium. What has been attributed to particle density is really magnetic density. But the photon at the

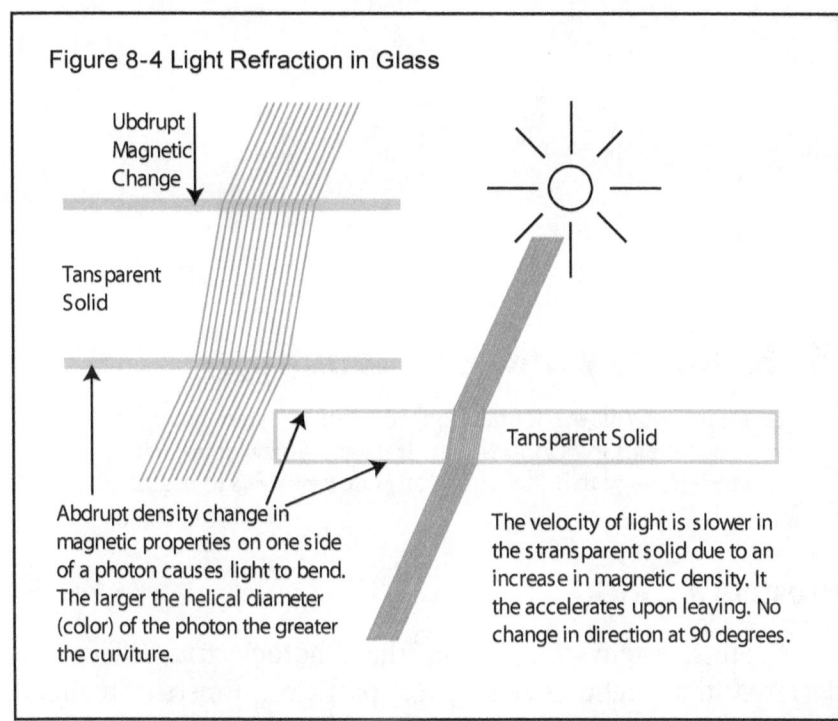

Figure 8-4 Light Refraction in Glass

Ubdrupt Magnetic Change

Tansparent Solid

Abdrupt density change in magnetic properties on one side of a photon causes light to bend. The larger the helical diameter (color) of the photon the greater the curviture.

Tansparent Solid

The velocity of light is slower in the stransparent solid due to an increase in magnetic density. It the accelerates upon leaving. No change in direction at 90 degrees.

opposite end, depending upon its size or color if you prefer, again bends into the more dense side of the field until the system gives up a bit of its momentum (and thus velocity) to the free particle of light. A vacuum may alter the atmospheric particle density, but not the magnetic field density created by the matter of the apparatus, with the earth and sun included. The velocity of light is not constant to the observer any more than a vacuum has the same magnetic field density everywhere in the universe.

The Special Theory of Relativity is only an illusion and under the above concept energy retains the classical definition of mass in apparent motion. Energy cannot exist of itself. Only with mass, energy becomes the means of accounting the predicate-action of reality.

9. Relativity and Epistemology

> Philosophy is the foundation of science; epistemology is the foundation of philosophy. It is with a new approach to epistemology that the rebirth of philosophy has to begin.
>
> Ayn Rand[1]

Einstein's Clock

In Einstein's paper on the Photoelectric Effect, he declared that light consisted of particles.[2] Einstein realized that the energy carried by a particle of light was proportional to its color frequency. Double the frequency and the energy doubles proportionally. The indirect proportion between color and energy gave the Danish atomic theorist Niels Bohr the clue needed in order to interpret the emission and absorption of light particles during the rearrangement of the more massive assortment of particles in the atom.[3] Because all the particles of an atom arrange themselves only in certain well-defined patterns, the exchange of photons demands this degree of order.   Nature produces atoms more faithfully than man produces a watch. Each kind of atom gives off light; like a fingerprint, a pattern of lines are generated. For the most part, light emits randomly, but Einstein reasoned that light particles of a particular color frequency could be provoked by the arrival in the vicinity of another particle of light of the same color.

> Put the vibrant particles in step with one another and they make a regular wave.[4]

In the 1960's, physicists succeeded in making laser lamps based on machine-gun-like movement of light particles.

This principle could produce light of extraordinary purity and coherence. The novel image-making technique of holography, also a by-product of provoking the regularity of light emission, brings us to the discussion of Einstein's clock.

The atomic clock came in a little earlier than the laser and became a practical timekeeper by 1955, the year of Einstein's death. Because of their accuracy they have become the basis of modern timekeeping. The standard atomic clock uses a continuous beam of atoms of the element caesium, each ready to make a certain rearrangement. In a cavity inside the clock, the atoms stimulate one another to make the change and produce a precise photon emission. Now the vibration produced by these atoms regulates a quartz crystal. This in turn drives an electronic digital display showing the time of day.

The regularly pooled readings from about eighty atomic clocks in government laboratories around the world end up at the International Time Bureau in Paris. The atomic clock has become more reliable as a timekeeper than the rotation of the Earth. In fact, geophysicists can study, with its aid, subtle changes in the length of the day. Atomic clocks can withstand heat, cold and hard blows. With the atomic clock, time measurement to the accuracy of a millionth of a second a day is routine.

Einstein predicted relativistic time under the General Theory. It has since been demonstrated simply by placing two different atomic clocks at different altitudes. Being synchronized before separation and later brought together for comparison after a period of time, the clock at the lowest elevation proves to run slower. Remember that as each atomic particle absorbs and emits photons at precisely defined color, the point should be made that the modern interpretation never considered that the reaction time between emission and absorption would change with a change in magnetic field density. Sorting out these influences proves a practical importance to the navigator using atomic clocks. A navigator, moving around within varying magnetic influences (gravitational to the relativist), accumulates his relativistic timekeeping errors. Unless he resynchronizes his clock frequently at known positions, these errors have known to grow and become significant as to atomic clocks.

Within a denser magnetic field, light photons move at a slower velocity than those in a weak field of a higher elevation. In other words, the velocity of a bundle of light defines relativistic time—the slower the velocity of the photon, relative to a specified field, the slower the clock.

> In 1971 two American physicists, J.C. Hafele and Richard Keating, carried out an experiment by taking portable caesium clocks (Four of them, for safety and reliability) right around the world in a passenger jet aircraft. They compared them at the beginning and end of the journeys with the reference clocks at the US Naval Observatory in Washington DC. One circumnavigation was made eastwards, with both journeys taking about three days. The result of the experiment indicated that the clocks no longer agreed about the time of day. The east bound clocks lost, on average, 59 nanoseconds (billionths of a second) compared with the clocks in Washington, while the westbound clocks gained 273 nanoseconds. [5]

How could such accurate timepieces disagree about the time of day when the velocities and altitude were relatively equal? Taking into account the General Theory of Relativity, the jet aircraft would place the atomic clocks in a gravitational or higher unified magnetic field weaker than the clocks at the Naval Observatory on the earth's surface. The clocks in a weaker field at the higher altitude would naturally run faster. This would explain the westbound clocks, for they gained 273 nanoseconds. But the eastbound clocks lost. Why such a difference? This cannot be dismissed as a Special Relativity or General Theory calculation. Instead, the net magnetic field density of the earth and sun will calculate to be greater in one direction than another. Consider that the earth rotates within its own magnetic field as well as the sun. This would suggest that a west-moving airline would be closer to rest relative to the net magnetic field while the earth rotates beneath. In the easterly direction the airline would move faster than the rotation and thus pass through many more lines of magnetic north to south motion than traveling west. The magnetic density would appear denser to the clocks and thus affect their motion. The high altitude would speed up the clocks, but the easterly motion would slow them down. That would explain the small difference. The high altitude with a more at-rest situation moving west would speed up the clocks by both situations. Let me explain further:

We must consider an observer on the earth to be a zero point of velocity calculation and understand that there is no negative velocity. Both directions are positive relative to the observer on the ground clocks. The Special Relativity principle is not the cause because both directions would produce a slowing of time. This leaves us with the relative magnetic field density. The earth rotates within the magnetic field in order for the earth to generate additional action on the magnetic axis. Although the magnetic field circulates vertically, it does not move rotationally relative to the earth. The observer point would then actually rotate through the magnetic field yielding an apparent increase in density. The jet moving in the easterly direction would add to that apparent density as one would put his hand out of a moving car window and feel the increase of the magnetic wind. In the westerly direction one's hand would almost sense a magnetic wind at rest because the jetliner is moving contrary to the rotation of the earth thus cancelling out an increase in magnetic drag. This is not electric particles that are sensed by the hand, it is magnetic particles that are sensed by the vibrations of the clock. The hand is only an analog for understanding the concept of a magnetic wind. The east-bound clocks would appear to move in a denser magnetic field and thus slow the clock workings down. The westerly direction would decrease the density even greater than the clocks moving with the earth—thus the difference between a loss of 59 and a gain of 273. If the magnetic field had nothing to do with the experiment, then both directions should produce the same increase because of the higher elevation and equal velocity.

Clocks go slower when at rest relative to any field, but move a clock through a field and it has the same effect as being at rest in a denser magnetic field. The velocity of light within an atomic clock depends upon the field density. The apparent density as determined by altitude, direction into the magnetic field and even the time of day—where the clocks at night would be further from the sun than during the day). The experiment was fallacious in proving relativity because these variables were not considered. Light travels slower in a gravitational field. And what has been understood to be the gravitational field that slows atomic clocks is more like a magnetic field density.

As a time clock runs slower in a magnetic field, it should be understood that the photo-electric fields propagate slower, not time. Time is constant. The velocity of light differs at different points in the universe. Even in a vacuum the earth's magnetic field density cannot be diminished. Time becomes an irreducible apparent reality distinguishable from the objective timepiece. The concept of time remains absolute by virtue of the law of conservation. This could be demonstrated if the angular velocity of a spiraling photon could be measured separately from the linear velocity. The constancy of motion lies in the angular velocity of light, regardless of the magnetic field density, but the linear velocity varies according to the magnetic field density.

The ideal limit of perfect time may never be known. We depend upon the measurement of objective reality as a close approximation of this ideal. Each measurement changes, but time will move on without a single care as to the accuracy of any clock. The idea of going forward in time or backwards is a fantasy and a total misinterpretation of the reality of time. We love fiction and fantasy more than conservation. Science should never have dabbled in such mysticism.

Angular Energy

In the objectivity of mass we were able to trace through the axiom of *distinction*, using *equality* in the balance scale and the law of *proportion*, to determine the nature of *conservation* in volumes and atomic mass. This process has given precise meaning to the terms used in physics. All of the above prior

Relativity and Epistemology | 166

axioms particularize the essence of *meaning* as to reality. We now turn to a special aspect of energy that is not taught in physics and also is not something even considered. It is a lot like trying to understand the meaning of things and the general concepts of reality rather than just learning how to predict the details and concluding falsely from them. It is apparent that more emphasis be placed upon the epistemology of meaning or the essence of understanding before we continue with the truth about modern relativity.

In the beginning of this thesis, the following axioms were put forth:

1 defines EQUALITY
2 defines DISTINCTION
3 defines PROPORTION
4 defines CONSERVATION

They are called axioms because of their precise meaning in that they cannot be altered. Axiomatic meanings corresponded precisely to their numerical values with the higher axiom providing the essence to the next axiom below in number and at the same time becoming a particular aspect to all of the higher axioms. Number five can now be added with this in mind. Five forms the basis of this chapter. It is the essence of conservation and each of the prior axioms form a particular part of this fifth axiom we might describe as self-evident *meaning*.

5 defines MEANING
6 defines MEASUREMENT

The sixth will also be added at this point. But first, *meaning* can rigidly incorporate all prior axiomatic terms. The particulars of meaning are: *One* demonstrates *equality* or a *synonym* that equals the term in question. *Two* demonstrates a *distinction* or *antonym* that reveals a difference or opposite term. *Three* demonstrates a *proportional relationship* that if associated triangularly with say a synonym and an antonym, the term can become even clearer. *Four* demonstrates *conservation*. The meaning of a term should maintain consistent or *proper usage*. Historical usage, as compared with a change in current usage,

The Einstein Illusion

illustrates times when the user neglects the fourth axiom when it comes to meaning. Such has been the case with space and time. *Five* is meaning itself—the essence of which is manifest by the *sixth* axiom referred to as measurement. A word's definition should be consistently generalized and regularly *contrasted* or *measured* against other words in order to maintain axiomatic qualities. Meaning is a particular part of the next axiom of *measurement* in that you cannot measure consistently and accurately without the clarity of terms maintaining their particular part of measurement. The meaning of inch, for example, must be conserved from one inch to another in order to maintain a quality measurement. As can be seen, all axioms of lower numbers form particular aspects of the higher.

The previous chapters have attempted to conserve and contrast the usage of words such as space and time. Mass and energy have also been conserved in meaning. The relativist's usage of these words has failed miserably. In essence, the usage of mass, time and space have been altered by modern physics in order to intimidate the conceptual mind to worship the mind that has no answer and prefers paradoxes. This removes responsibility to reality and places in its stead priests who are exalted for keeping humankind in darkness, away from a universe that manifests the eternal consistency of God. Meaning constitutes the foundation of philosophy and lies halfway between the unity of *one* and the perfection of *ten*. Meaning represents the axis of all reality and more particularly the axis of truth.

What are the particulars of measurement? Take the five senses, but please include the sixth as intuition and the result is six aspects of measurement. The senses may be too human to be considered a very precise form of measurement, but the general idea that observed fact is considered the basis of all experimentation, it might then be well to accept the senses as measurement. In physics, we need to go a little further not forgetting *equality*, *distinction*, *proportion*, *conservation* and precise *meaning* of terms. All of these particulars are found in six sections of direct proportional equations that help us measure physical reality. Each section has three basic equations with the law of conservation manifest in a progressive derivative of each section. Proportion is inherent in each equation and distinction is found in the relationship of the three basic realities. Of course, equality is what each equation seeks. The fourth equation in each section is generally not used, but gives a glimpse into the seventh axiom which will be discussed later. The fourth equation in each of the six equations of measurement are more intuitive than the prior three and require greater perception as would the sixth sense.

BASIC EQUATIONS THAT MEASURE PHYSICAL REALITY

I. Linear Position
   <u>Position = L</u>
   Velocity = L/T
   Acceleration = L/T^2
   Control = L/T^3

II. Angular Position
   <u>Radial or Arc/Radius = R</u>
   Angular Velocity = R/T
   Angular Acceleration = R/T^2
   Angular Control = R/T^3

III. Linear Position of Inertia
   Moment of Mass = $M \times L$
   <u>Momentum = $M \times L/T$</u>
   Force = $M \times L/T^2$
   Mass Control = $M \times L/T^3$

IV. Angular Position of Inertia
   Moment of Inertia = $M \times R^2$
   <u>Angular Momentum = $M \times R^2/T$</u>
   Moment of Force = $M \times R^2/T^2$
   Angular Mass Control = $M \times R^2/T^3$

V. Linear Kinetic Inertia
   Kinetic Inertia = $1/2\ M \times L^2$
   Kinetic Action = $1/2\ M \times L^2/T$
   <u>Kinetic Energy = $1/2\ M \times L^2/T^2$</u>
   Kinetic Power = $1/2\ M \times L^2/T^3$

VI. Angular Inertia
   Angular Inertia = $M \times L^2$
   Action = $M \times L^2/T$

Energy = M x L²/T²
Power = M x L²/T³

Section I lists one equation defining *position* and three equations defining the *rate of change in position*. Control, the third derivative, explains that:

> To increase the speed of an automobile, we push on the accelerator, causing positive acceleration. To decrease the speed, we step on the brake, causing negative acceleration. We may also alter the direction of the car by steering. What is the process by which the accelerator brakes, and steering changes the acceleration of the car? Clearly, a change in acceleration is what we mean by the word *control*.[6]

The fourth equation in each section reveals the measurement of perhaps intelligent choice or control value.

Section II lists one equation that defines a *radian* as a length equal to the radius and three derivatives showing the *rate of change in angular position about an axis*.

Section III lists the first equation as defining the *moment of mass of a particle* or *position of inertia*. If *m* equals the mass of a particle and *l* its distance from a point called the center of mass of the system, the moment of this distant particle of mass or position of inertia equals the product of *ml*. In other words, if the portion of mass or particle moved to the center of mass of the system, the moment of mass of that particle would be zero. All mass particles participate as part of a system, but if the particle moves along an incline plane, the equations speak of linear position of inertia and the three equations that follow define the *rate of change in the linear position of inertia of a single mass particle*.

Section IV first lists the angular position of inertia when the moment of mass is greater than zero. The *moment of inertia* is greatest when the total mass of the system lies at the parameter. Thus gyroscopes and fly-wheels have the greatest moment of inertia. Angular momentum and torque define the first and second derivatives of the moment of inertia represent the *rate of change in the angular position of inertia*. Angular mass control is added as a third derivative and describes the rate of change in the torque or moment of force. If the mass distributes evenly from the center outward, the average moment of inertia, or the angular momentum or torque of each particle, equals the total of the system.

Section V lists the various derivatives of linear or *kinetic inertia*. Kinetic inertia derives backwards from kinetic energy. Inertia comes from the Latin word meaning "idleness," or a kind of unwillingness to make a change in direction or position. The condition that mass has in resisting change may be regarded as one of Newton's laws. For this reason, Newton's third law states that for every action an equal and opposite reaction will result. The *1/2* in each of the kinetic equations can simply define the amount of inertia possessed by the linear moving mass as one half of the total inertia in the system. Action psychologically describes one half and the reaction the other half of the total participating result of the motion. A simple example is that a bullet fired from a gun only carries one half of the total potential energy before firing. The gun recoils with the other half—the other one-half moves in the opposite direction as a result of the release of the total potential energy. Kinetic action, kinetic energy and kinetic power define the first, second and third derivatives of the *rate of change in inertia with a change in linear position*.

Section VI lists the equations for angular or total inertia. Angular inertia here defined has little meaning to the textbook reader as in some of the other third derivatives. The essence of angular inertia here lies in the fact that it has the same analysis as the angular position of inertia. For this reason, energy can be defined as a derivative of angular momentum. The only difference between angular position of inertia and just angular inertia lies in the difference between vector and scalar quantities. The difference between the moment of force and energy implies just that. One denotes a vector quantity and the other denotes a scalar quantity. Both have the same elements. When considering the total system, the quantity amounts to a scalar quantity under angular inertia. When only a portion of the mass (as in the angular position of inertia relative to a center of rotation) is considered, a vector quantity results. The third derivative, or *power*, yields the rate of change in energy. If one worker can perform more work than another in the same period of time, that person has more power. Energy equals the total work done. Power equals the amount of energy released per second of time.

Each section alternates between linear and angular equations. Each equation represents the first state of inertia and three rates of change or derivatives. The physical scientist or mathematician endeavors to find the rate of change of either position, point of inertia or inertia itself. Conservation equations, underlined in each section, simply progress to a higher derivative in each section pair. Interestingly, equality imparts conservation to position and equality coupled with distinction imparts conservation to a radial. A perfect proportion applies to momentum, and in addition to the prior three, conservation itself emphasizes angular momentum. Conservation of kinetic energy matures in the meaning of "one-half" and energy itself reveals the ultimate in scalar measurement. Measurement gives the ultimate process for intelligence to measure reality in terms of subjective space, predicate time and objective mass. Epistemology requires that the meaning of terms remain conserved. Relativity, on the other hand, turns meaning upside down, diminishing the intelligence of clear meaning by destroying the conservation required to keep the meaning of reality.

The whole essence in understanding the development of the equations of physical measurement requires that one understand the difference between the linear and the angular aspects. Psychologically it is the difference between the whole system and its radiant parts. The whole system, such as this book you are reading, has a zero vector direction of motion relative to you the reader. But consider that any particle therein and its angular motion begins to yield a sort of angular direction. The only way to measure this book as to total inertia comes by measuring it as a scalar total of individual vector units having the same angular position of inertia. Simply measure the torque of a single unit, multiply that by the number of units, and the scalar result will be the total potential energy of the book. In a more complicated or directional fashion, imagine that each and every photon particle held by the angular momentum of the magnetic field, one at a time, exploded or spiraled off from this book by the conversion of angular motion to linear motion in the spiraling photon. Measure the linear kinetic energy of each photon spiral, and then add to it the linear kinetic energy of the book, which would eventually be zero. Realistically, simply weight the book and multiply it by the velocity of light squared. This implies the same as measuring each and every photon as it accelerates. If the circular velocity of the magnetic field, holding the photo-electric field at rest, equals the linear velocity of light, does not the equation of energy $E=mc^2$ say exactly this—that mass is angular energy and light is linear energy? It is the six equation categories that give us this clear relationship, but modern physics ignores angular energy even though proportion, conservation, clear meaning and measurement suggest it.

The kinetic energy of a single particle is equal to one-half the total energy expended to yield the linear path of travel. A particle possesses a moment of force or angular energy when at rest linearly. Here called angular or potential energy, relativity names it poorly as "rest energy." Understanding more about vector sums and vector products can reveal the importance of understanding why the velocity is squared. But this has nothing to do with a second dimension. All derivatives take place on one dimension or direction. For each direction, such as in two or more forces having different directions, the second derivative pertains to each. Gravity, for example, has one vertical direction and a billiard moving across the table has another horizontal direction. Each force applied passes energy into one direction. Thus in each equation, using *velocity²*, you square the numbers and not the units such as meter or inch. Given this v^2 ignores direction. The term still remains as a single direction.

Figure 9-1 demonstrates a closer look at a linear vector using a mass with two forces applied by two external particles not

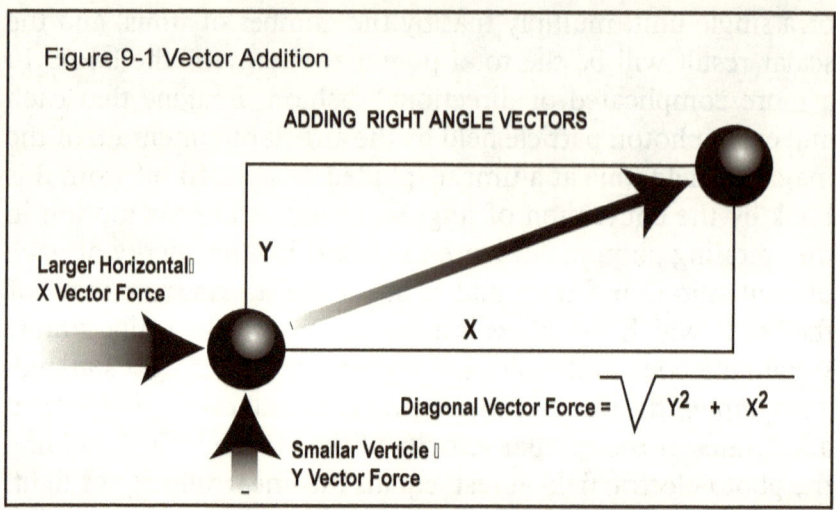

Figure 9-1 Vector Addition

shown. The two objects applying the two forces will each receive a force equal in magnitude but opposite in direction to the forces they apply. The net result of the illustrated particle then takes up one-half the total linear energy of the system and moves with a vector force equal to the diagonal of a parallelogram, if the two forces applied can be represented by the two sides. The two forces cannot be added to give the total force because the sum of the two sides of a parallelogram is greater than the diagonal. The square root of the sum of the square of the two sides (two forces) must be added to equal the diagonal (net force). Velocity and acceleration work the same way. This is called *vector addition*. Momentum and energy, on the other hand, must be added in the *scalar* fashion. Momentum and energy are conserved, velocity and acceleration are not.

Figure 9-2 demonstrates an angular vector using a mass entering a greater field density. The particle in question moves counter-clockwise. Using the linear vector method, the direction of the particle can only be calculated for small segments or tangents to the curve, thus treating the curve as a straight line. With an angular vector, however, one vector comes from the particles straight line inertia or momentum and the other vector lies at right angles to the particles' motion and results from a change in field density perpendicular towards the center of the system. The two vectors reveal the essence of *vector square* or

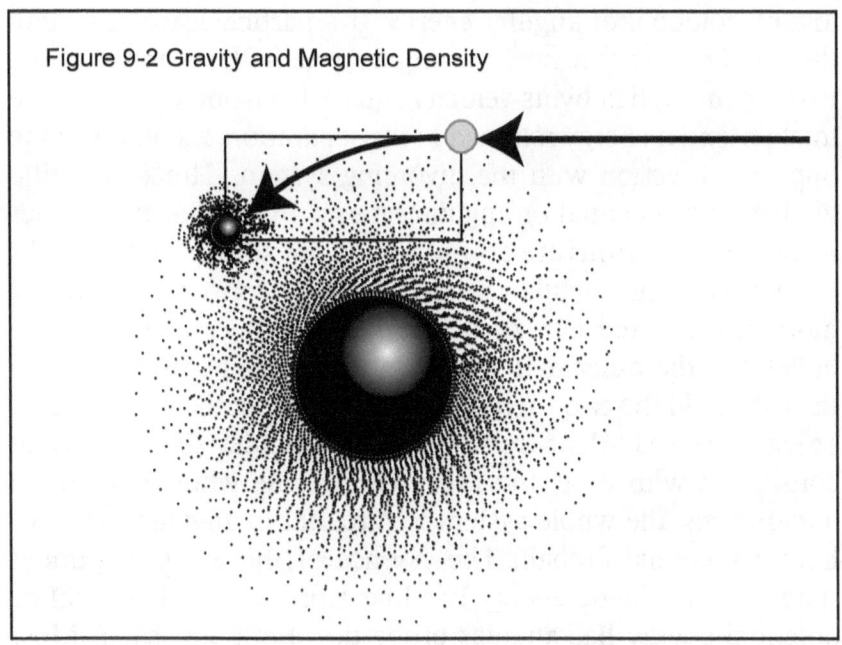

Figure 9-2 Gravity and Magnetic Density

velocity square in Einstein's $E=mc^2$. The net result (linear vector multiplied by the radius) or vector product yields a curve similar to the diagonal in a straight-line vector.

Since the particle entered an orbit due to the greater field density, one half of the linear kinetic energy converts in part to the energy required to maintain the radius while the other half remains as inertia to the particle in orbit. Although the new orbiting particle still has apparent kinetic motion, the whole system measures zero linear kinetic energy because the whole system has a center of mass that remains stationary. The total energy is potential. Kinetic energy or $1/2\ mv^2$ can only be measured in the individual linear motion of the parts, but treat the whole system as a single unit and the linear energy now equals zero, thus the term "rest energy" when linear motion is not apparent. But consider the individual photons orbiting at such a velocity that if released with the same velocity upon entry, the *angular or potential energy* $= mv^2$.

In an angular system the total energy remains constant as long as the mass remains constant. Since mass is proportional to the resistance to change, then a loss of mass also accompanies a

The Einstein Illusion

loss of potential or angular energy. If a particle leaves a system the kinetic energy it now has equals one half the mass of the particle multiplied by its velocity squared. The other half of the total potential energy released in the separation, rebounds in the opposite direction with the divorcing system. Think of a rifle loaded with potential or angular energy. In fact the gunpowder is just that. Electrons are orbiting ready to explode. With a slight disturbance, the angular motion will be converted to linear motion. Once the gun fires, one half becomes kinetic to the bullet and the other half moves in the opposite direction with the rifle. Add the two together and it will equal the total angular energy released. Heat from moving molecules should also be considered with each and every part of the whole system. In linear terms, the whole system is zero, but in angular terms we have a potential fireball. The essence of the whole argument states that if kinetic energy has linear properties; then total or potential energy has angular properties. Both are scalar. Most important, one is converted to the other. Simply put, angular mass is converted to linear mass in the same way as angular energy is converted to linear energy. We can skip angular and linear mass all together by including mass in energy and saying that energy is mass in motion.

    The idea that the kinetic energy of a particle equals one half the action of the total energy released seems perhaps more psychological, but analysis defines it to be mathematical. If it can be concluded that energy manifests action in linear or angular conditions and if the mass that carries the energy moves linearly or angularly, then spinning electrons, protons and neutrons have angular or potential energy, if considered systems with a center of a rotating mass. Potential or angular energy applies only to the whole system in rotation about an axis. Consider also that, according to previous generalizations, each of these particles composes a large number of photo-electric particles (photon spirals). Disturb their magnetic equilibrium and the angular momentum of the particles' system converts to linear momentum for each photon bundle of light released. This epistemologically provides another way of grasping the meaning of $E=mc^2$. This provides the assumption that each and every photon has the same angular momentum, whether spiraling around a magnetic field in motion within the angular system or spiraling around a magnetic field at rest when moving in a linear direction away from the system.

The Sun compresses matter through the use of dense gravitational (magnetic) fields. This pressure breaks the nuclear equilibrium and releases a substantial part of the angular energy of matter. Relativity calculations for such a feat assume that matter is at rest. Little be it known that each and every particle carries a whirlwind of "angular" momentum. The word "rest" seems proper only because the particles' center of mass stays at rest relative to the linear measurement of light. Rest pertains to linear motion of the system and not to the angular properties within.

Most of the visible stars in the sky move in mid-career, steadily converting their angular momentum and energy, or as some prefer, steadily burning their nuclear fuel. Mass and the energy that mass carries cannot be created or destroyed in a closed system, so the missing part of mass as potential or angular energy reappears as both linear momentum and kinetic energy. Mass and energy are not exchangeable in the way that one can convert dollars into gold or vice versa. They always remain the same thing, mass-energy together. Epistemologically this should mean 'mass *with* energy'.

> In the Cambridge atom-splitting experiment the flying helium fragments together possess, for a moment at least, exactly the same mass as the combined mass of particles that produced them.[7]

This implies a conservation of mass, but where did the energy of motion come from? Obviously from the angular energy already present within the very dense nuclear magnetic fields. Under this epistemological definition of $E=mc^2$, both mass and energy must be conserved in nuclear reactions.

A relativist prefers "rest energy" for conventional mass. Proper for measurement and the calculation of kinetic energy, but because the total linear momentum and kinetic energy equals zero, the rationalist prefers to understand "angular energy" within any mass.

To illustrate the attitude of the typical theoretical physicist Richard P. Feynman, the Nobel Prize winner told of a personal story about light out of matter:

> Once, though, when I came back from MIT (I'd been there a few years), he (my father) said to me, "Now that you've become educated about these things, there's one question I've always had that I've never understood very well."
>
> I asked him what it was.
>
> He said, "I understand that when an atom makes a transition from one state to another, it emits a particle of light called a photon."
>
> "That's right." I said.
>
> He says, "Is the photon in the atom ahead of time?"
>
> "No, there's no photon beforehand."
>
> "Well," he says, "where does it come from, then? How does it come out?"
>
> I tried to explain it to him – that photon numbers aren't conserved; they're just created by the motion of the electron – but I couldn't explain it very well. I said, "It's like the sound that I'm making now: it wasn't in me before."
>
> He was not satisfied with me in that respect. I was never able to explain any of the things that he didn't understand. So he was unsuccessful: he sent me to all these universities in order to find out those things, and he never did find out.[8]

Feynman had many stories from his life that taught him never to trust things that he did not understand. But in the case of relativity physics he chose the mystical path as all have been conditioned to since the contemporaries of Einstein began coining the new terms. His father did not accept the idea that light was not in the atom in the first place. To say that sound was never in the mouth is an escape from responsibility. Sound waves and photons are two completely different concepts of energy. Sound is like the billiards exchanging momentum from one to another as illustrated previously, but light is like a single field traveling the whole distance. It is not a wave in a medium in traditional terms. Rather it is one field moving through the medium of another. Relativists cannot explain it because they do not understand. It was not Feynman's father that was at fault. It was Feynman himself for being intimidated by the traditional darkness of modern relativity.

Relativists do not look at the mathematical equations closely. They too often step out of the equation and conclude a change in terms as to their reality. It is like saying that substance is converted to motion without substance. It does not matter how educated a person is, tradition is stronger. We need to learn the difference. It is like following the rhetoric of tradition rather than the logic of reason. We all do it more than we care to admit. Light is mass in motion and there is no getting around it.

> The Earth intercepts very little of the Sun's out pouring, yet about 160 tons of sunlight particles fall on the Earth every day.[9]

Green plants absorb the Sun's linear motion of light particles into angular motion of the magnetic field to build carbohydrates out of water and carbon dioxide gas. As you read these words the retina of the eye converts the linear momentum of every photon into angular momentum. The atoms then exchange photons and electron particles from one atom to the next until a single atom in the brain vibrates when it receives the momentum of the last photon. The last atom in a chain can bind the constituents of matter in such a way that the final absorption releases a very low energy photon in the infrared range. This conclusion is based upon the concept that electric current is light itself and not the motion of electrons.

The microcosmic universe continues to jostle around giving up and absorbing each particle and its accompanying energy from the angular momentum to linear motion and back again. Mathematically, angular energy seems meaningless, but geometrically it describes the energy possessed in two moving fields that I call the photo-electric and magnetic fields. Both fields have substance and motion. Both form the creation of mass and light.

Solving for Mass

The point of this section is to discuss the modern relativist's claim that mass puts on weight at higher velocities. This would not be offensive if light is added, but to expect mass to increase without adding light is fallacious. By adding light is meant absorption of the linear velocity of a photon into angular velocity in the mass. The mass of the photon is added.

The Special Relativity principle appeared on the scene in an effort to retain the laws of conservation assuming the constant velocity of light, therefore mass increases at high velocities to algebraically account for relativity's claim as to a change of space and time. It would be impossible to take a moving particle and weigh it on a stationary scale, so the physicists' method of determining mass relativity comes from indirect means such as particle refraction in a magnetic field. If a charged particle moves within a magnetic field, it will curve. The deflection varies depending on the charge of the particle, its mass, velocity, and the strength of the field. Before the discussion of particle curvature in a magnetic field, it helps to understand how the particle reaches such high velocities in order to even consider such things as putting on mass.

Figure 9-3 demonstrates how charged particles enter a long, evacuated tube within which lie a number of hollow cylinders of increasing length.[10] The cylinders connect alternately to opposite terminals of an electric field generator, as positive ions drift through each cylinder at an accelerated velocity. When the ions pass into the gap between the first (positive) and second (negative), the positive ions are accelerated forward by the electric field within the wiring. The positive ion moves away from the positive cylinder and towards the next negative. It is actually the magnetic field that does the pushing while the electric current pushes the magnetic field. Upon leaving the second cylinder, the electric field (current) reverses on all cylinders by means of a radio-frequency generator. The negative cylinders surrounding the positive ions become positive and kick the positive ions onto the reversal of the electric field of the next cylinder as it becomes negative. Whenever a charged particle moves in the interior of a cylinder, a shield prevents any effect during the electric field reversal. The field generator alternates at a constant frequency, thus the cylinders increase in length as the particle accelerates to be in sync with the acceleration of the ion.

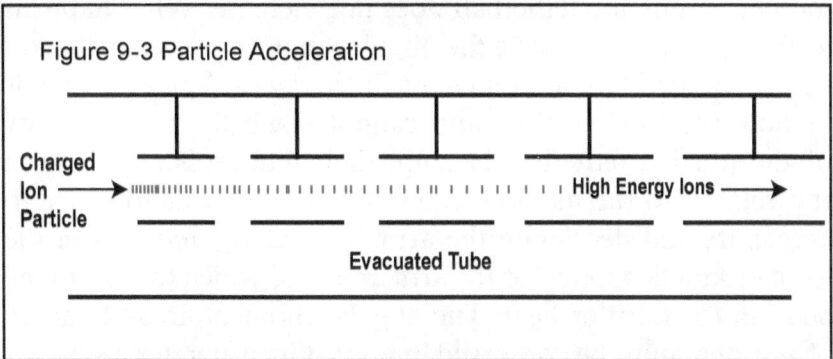

Figure 9-3 Particle Acceleration

The relativity principle assumes that an increase in the energy of the electric field in some way transfers to the increase kinetic energy of the moving particle. But the particle does not accelerate sufficiently. Where did the energy go? According to Einstein's $E=mc^2$, the energy expended does not kick the particle to a higher velocity as it should. The extra energy appears into what relativity calls an increase in mass of the particle. This calculation solves for and conserves energy by making mass and energy convertible. The law of conservation demands an explanation and this is what the relativist has come up with. It may not be easy to understand fully how an electric field passes energy, but a mechanical example may help explain the assumption.

Imagine two tether ball players and, for fun, let us change the game. The opposing player, instead of hitting the ball in the reverse direction, attempts to increase the angular velocity of the ball around the pole. The player's arm becomes the analog of the electric field within each cylinder and the tether ball mimics the ion. Whether the field or a hand gives the kick, it can be assumed to be an expenditure of energy. This analogy is comparable to a cyclotron particle accelerator and would easily gain a substantial speed if skilled players had the capacity to increase the angular velocity of the tetherball sufficiently (assuming the rope was rotatable without friction). At greater velocities, it becomes difficult to further increase the speed. Once the tetherball reaches a high angular velocity and despite the increase in energy expenditure of each player, the angular momentum of the tetherball does not increase. What happens to the energy? Obviously, the angular motion of the arm takes up the energy. No matter how much energy each player expends or how they swing, the hand cannot reach the high velocity of the moving tetherball because their motion is angularly at rest relative to the moving tether ball. They constantly have to accelerate and decelerate the arm. The energy remains in the form of kinetic motion of the arm, much of which returns to the body in the form of heat. The angular momentum and energy of a system must be conserved, but what incorporates a system? In this case it is the tetherball and players as a whole.

In the particle accelerator, the energy in the system remains in the kinetic motion of not only the ion but includes every electric and magnetic field. It does not altogether pass to the particle unless you can be assured that both the electric and magnetic fields can equally accelerate with the charged particle. The physicist may calculate the lost energy as kinetic energy remaining in the electric and magnetic fields—much of which may end up as radiant heat that leaves the system over time as very low frequency photons. This will be explained more fully in the discussion of black-body radiation. The action, however, in the electric field that drives the magnetic field which in turn drives the particle directly produces no means of direct measurement save the electric field. If the electric field has a measured momentum, the momentum of the magnetic field must also receive the same momentum including a reversal of motion. If the electric field is a constant alternating current, the field has to stop and reverse direction in much the same way the tetherball arm has to stop and accelerate with each motion. This stopping and starting makes it impossible for the fields to increase in velocity proportional to the velocity of the charged particle.

Under the theory of relativity, the mass of the particle admittedly would be the same if the tetherball players rode along at the speed of the particle. But if each tetherball player was able to accelerate with the tetherball, each could skillfully accelerate the tetherball because the ball would be more at rest relative to each player moving also. But remember it would take extra energy to accelerate each player. Do not assume that this expenditure of energy to accelerate each player will move onto the ball any more than the energy accelerating an electric field cylinder would pass to the particle.

A general point to make is that Faraday believed that both the magnetic field and the electric field had particles pertaining to each. This implies density as well as strength in terms of motion. Maxwell put Faraday's experiments into mathematical form by giving the fields quanta measurements but somewhat neglected the existence of magnetic particles. This suggested that energy was pure motion without the movement of a mass of particles that create the field. Relativity assumed the same and thus denies the existence of any matter field that holds the energy. If the magnetic field can really hold energy, relativity can certainly be questioned.

When physicists measure the curvature of a charged particle passing through a magnetic field, the curvature depicts the total mass of the particle. If the particle does not curve, as it should, assumption says, "It puts on weight." But how much time does it take for an electric field, embodied within the moving particle, to be affected by a magnetic field that remains stationary relative to the direction of travel?" If the time is greater than at rest, then the faster the electric particle moves through a magnetic field (at rest vertically to the observer), the less action the magnetic field will have upon the charged particle. Thus high-speed particles may curve less due to the fact that the magnetic field does not have time to pass momentum to the charged particle. Just as in the tetherball players, all the energy never passes to the particle. The particles avoid the same curvature as at rest because when the particle passes through as fast as the motion of the field, there is no time to react. There are other reasons, but they will be discussed later. It has to do with the polarization of any particle.

The scientist feels certain about the immutable nature of conservation laws especially of energy, but he speculates upon the nature of field physics of which the equations of relativity are dependent. The Special Theory of Relativity can be derived by the Lorentz transformation assuming the constant velocity of light, but Einstein arrived at it differently. Einstein's Special Relativity Principle originated from Maxwell's field equations which then again were based upon Faraday's particle fields. If the field is a whirl of particles, as Faraday suggested and as Maxwell seemed to have put in mathematical form, it would not be difficult to pass energy from substance to substance, but to pass momentum and energy upon a particle moving faster than the motion of the fields poses a problem because it takes so much energy to stop and go a field before it even has a chance to react with something that moves faster and faster. The faster and faster the particle moves something that has to stop and start cannot pass energy anymore than the tetherball players.

Relativity collects a mass of assumptions mixing up reality like a cup of soup. It might even use equations based upon a field theory, but denies the field as a physical property. Knowing that both the electric field and the magnetic field can have physical properties, Newton would have modified his equations to the extent that gravity is not infinite and Einstein would have corrected his epistemology before it was too late. This is the essence of the problem as we continue.

Relativity is not a byproduct of mathematics. It is a consequence of changing the concepts of the terms in the equations. Relativity also changes the implication of the operators. In order to times something with another quantity relativity forgets if one term is zero the whole is zero. You must have zero energy with zero mass. You can assign energy an infinite velocity and think you have something, but when the mass is zero the energy is nothing. Velocity is also nothing. How then can you convert mass to energy? Energy is the velocity of mass. Take away the mass and the energy including velocity is nothing. Conversion is angular momentum to linear momentum. Both have mass, both have velocity and both have energy.

Samuel Dael

10. Action At A Distance

At first sight, gravity, electromagnetism, the weak, and the strong (nuclear) force have nothing to do with one another. But as physicists explored deeper into the structure of matter, the distinction between the interactions becomes superficial. That is the major message of theoretical physics in the last decade.

Heinz R. Pagels[1]

Particle or Field

Careful observation leads us to the conclusion that something invisible happens with many forces in nature. On the microcosmic level, atoms pitch and catch electrons while electrons exchange photons. When these point-like particles curve like microcosmic planets with certain attractive or repelling forces, the physicist talks of a force called charge. On the macrocosmic level, gravity demonstrates the most obvious force in nature and the thinking man has come to accept the earth's larger mass as a direct cause of attraction. With nothing better to conclude, Newton moved inevitably to the concept of *action at a distance*. Like magic, one mass affects another through empty space. Interestingly, the net acceleration of gravity in this universe seems never to allow contact between star and planet. Action at a distance is not something that fits reason because one would think things would collide. Things do not seem to react to other things at great distances. We only assume that what the eye sees is sufficient for a mystical cause. If there is no such thing as empty space, things fall because space is not empty. Something local is causing the movement or the acceleration.

Physicists talk of four basic forces in nature. The weakest force hinges upon the gravitational interaction. It binds the planets. The electromagnetic force binds the atoms and creates the most obvious field pattern in nature. The strong interaction or nuclear force exhibits its effects at the rate of 100 times the strength of the electric force. It binds the atomic nuclei. Some writers have talked of two weak forces and an additional color force called the chromo dynamic interaction between the basic units of the proton and neutron called quarks. Though week at short ranges, this color force exhibits a very strong reaction.[2] Quanta, called gluons, mediate every interaction. The gluon of electromagnetism is the photon, of the weak interaction we find the weak gluon and of the strong quark-binding force in the atom's nuclei we find color gluons. Named the graviton, the gluon of gravity remains undiscovered. Today physicists hold that all interactions become unified at ultra-high energies.[3] The unification of all interacting forces was a dream of Einstein and an adventurous undertaking until his death. There is an apparent strangeness about gravity, but the attempt here will be to unify all of the forces of nature as a single electromotive force between the photo-electric field (light that never rests in terms of angular momentum) and the magnetic field (magnetic space that is both dynamic and at rest).

A unified field theory stands today as a very complicated structure. Nonetheless, the field concept used in electromagnetic interactions can also be applied to gravitation. Einstein's intention was just this, but the idea of a straight line vertical gravitational field is the problem.

> The results would do away with the Newtonian concept of gravity as a force acting at a distance, a notion Newton himself had admitted was inexplicable. "Action at a distance," Einstein wrote, "is replaced by the field."[4]

Particle physicists have referred to the gluons as the only physical reality for a field concept. If a photon is a gluon, then photons harnessed within a wire become the electric field. If the photon and its gluon nature skip through open magnetic space as an electric spiral helix having intrinsic inertia, the geometrical center of the electric spiral carries the center of momentum or mass of each gluon. It reacts to the magnetic space at rest. If a particle defines a stationary photo-electric field with its own circulating magnetic field, the geometrical center of both the electric and magnetic field equals the center of mass of that particle. Epistemology echoes again: Fields cannot be pure energy; rather they are substance or inertia in motion. Electric fields are gravitational, but magnetic fields are not, but that does not mean that the magnetic field does not have inertia as in magnetic substance different from electric substance. This limits the field concept to only two inertial fields. Coming up with several gluons may only say that they are just different photons. Every photo-electric field within a particle reacts to magnetic space of relative density vertical to the earth. You do not need a gluon of gravity when every photo-electric field inside mater is a gluon of light at rest linearly but not angularly.

Fields cannot simply be a strange form of energy in empty space without being an inertial substance; rather fields are different mediums (a massive number of minute particles) in relative motion that carry energy like any particle carries momentum. Mass does not fall as in action at a distance. It is the electric fields in mass that react to gradual change in magnetic density. Even if the density change is ever so slight, the extremely high angular velocity of each photo-electric field will more than account for something almost immeasurable.

> The quantum theory describes the interaction of sub atomic particles through the field concept. At first it seems that a particle has nothing to do with a field like a magnetic field, but as we describe the field concept it should become clear that particle and field are complementary manifestations of the same thing...
>
> In the Old Newtonian theory of gravity, the actual existence of the gravity field was not required and it did not have any material reality. It was simply a useful mathematical fiction for describing the effect of gravity on particles of matter. One could describe gravity as well without it.

The field concepts—that fields have a physical existence——came into its own in the nineteenth century. Michael Faraday the English physicist, who did extensive experiments on electricity and magnetism, especially emphasized the physical nature of the electric and magnetic fields. Electrically charged particles were viewed by him as points at which the field became infinitely large—the field, he argued, was the essential physical object, not the particle. Faraday's intuition about the physical nature of the field was finally realized in James Maxwell's electromagnetic theory of light. Electric and magnetic fields were not mathematical fictions in Maxwell's theory but could carry energy and momentum. Fields had a physical reality... But light also carries momentum and exerts a 'radiation pressure'—a small but observable push.[5]

Depending upon the author and especially if one is talking about relativity, he will conclude that Maxwell did not feel that the field had substance. This comes from the fact that before the advent of modern quantum field theory, physicists thought of particles and fields as distinct entities. Particles, considered to be immutable, were eternal while energy emanated from particles in the form of a field. The fields became responsible for the forces. Confusion was compounded when relativity became fixed on this mass-less energy, like ghosts having power to move things. How can energy leave a particle and become a field without taking a piece of both the action and the mass? Large particles are the geometry of various fields and should not have been considered to be permanent solid-like marbles. Also, a field should have been considered a compilation of the smallest units of physical reality. Electrons and Protons as a solid like substance is a misnomer. These particles are nothing but field patterns and the geometry of differing field concepts. Particles often behave like wave fields and wave fields often behave like particles. It makes perfectly good sense when the field structures substance rather than the field being something that emanates. Also, it makes perfectly good sense that the field, like a particle's outstretched arms, interacts in varying degrees with the magnetic space extended by other matter.

Quantum field theory suggests that everything can be explained by the interaction of two quantum fields and the flux density of the magnetic field provides the statistical probability for finding an electric particle field. The difference between a composite particle and its component field lies in the measurement of large statistical numbers as having a center of mass. The modern quantum field equations may be correct but the popular epistemology of the word *energy* in terms of fields without substance yields a paradox. The magnetic field does not emanate. It circulates with specific momentum. If the electric particle were to disappear, the magnetic field would simply disseminate its motion. I hold Philosophy responsible because it has failed to maintain the epistemology between the objective reality of mass and the predicative reality of energy—between substance and the motion of substance. You do not convert substance to motion; rather you convert one angular motion of substance to a linear motion of substance or the reverse. Energy does not emanate from substance unless it is a piece of that substance. Energy is intrinsic and is always there as a field having momentum. Momentum means inertia and inertia means some form of substance.

The first step toward a unified field theory begins by assigning energy to the action of substance and not the conversion from substance. By replacing action at a distance with an objective magnetic field composed of magnetic monopoles and a photo-electric field having its own particle, field theory is manifest.

> Hypothetical magnetic particles...have been postulated on conservation and symmetry principles: an electric particle gives rise to an electric field and when set into motion gives rise to a magnetic field; a magnetic particle should give rise to a magnetic field and, in motion produce an electric field. Neither quantum theory nor classical electromagnetic theory bars the existence of the magnetic monopole but it would have profound effects on the theoretical basis of quantum electrodynamics. Maxwell's equations would prove completely symmetrical if such particles did exist.[6]

The equations would be symmetrical in small situations, but at infinite distances magnetic space is apt to be more or less at rest and less dense far from the confines of matter. In this respect you could not project the velocity of light infinitely in all directions as most have done. Light would have to vary according the makeup of magnetic space. If light varies according to electric density, why does it not vary according to magnetic density? Was it not the magnetic field that really determines the velocity of light Since quantum electrodynamics is a relativistic theory of quantum mechanics? Relativity bars the existence of an indivisible magnetic particle as a constituent part of objective reality because of the meaning of the terms used in the equations. Relativity denies the indivisible, objective existence of both photo-electric and magnetic monopoles. Modern relativity holds on to its mystical form of energy and the constant velocity of light. Both are fallacious.

Accept Faraday's original theory that fields are a collection of electric and magnetic monopoles and field theory has its origin. Relativity in terms of the constant velocity of light cannot be proved by Maxwell's equations because the fields in question are particles having a variable density and are not energy that emanates from particles. In addition to this, what we measure as gravitational mass is the interactive nature between the two particle fields. What the effect a change in density of magnetic monopoles has on a field of electric monopoles is in question. It you change one field, you of necessity change the behavior of other. Finally, what we know as inertial mass is the number or quantity of photo-electric monopoles. It takes a certain number to make the smallest electron and perhaps a single photo-electric monopole is the smallest quantum photon. More germane to the issue is that electrons, protons and neutrons are not particles, but field combinations of photo-electric monopoles held in close proximity by circulating magnetic monopoles— the greater the numbers of photo-electric monopoles the more magnetic monopoles are required to hold the body together or at rest relative to the magnetic axis. There are no particles; there are only fields of two basic units. This does not, however, bar from existence a third undetermined monopole. Our fingers can touch the photo-electric field and sense it impinging upon our skin or retina of the eye, but the magnetic monopoles will pass through any attempt to discover them directly. In this respect they are different and would suggest that one cannot be converted to another. In other words, all things are not one. All things are also not many, otherwise we must break things down. Intuition suggests than the ideal number would be three basic fields. I will suggest this case at the end of this section.

Field Geometry

Now If the photon spirals through space carrying a bundle of photo-electric monopoles we attribute as an electric field and if this photo-electric spiral comes to be absorbed by an electron in need, then the electron can be defined similar to several electric fields stacked vertically in the shape and size of an electron.

If the electron obeys the right hand rule, the thumb pointing up describes the direction the magnetic monopoles move through the axis of the electron. The fingers describe the direction of the photo-electric particles as they create the angular momentum equal to the axis momentum of the magnetic flux. As shown in **Figure 10-1**, the geometry of the two fields yields a form of an objective electron with a center of mass at rest relative to both fields. What can actually be observed are only the photo-electric fields and not the magnetic portion because excessive photons bounce off the electron or other complex field creations, but pass through magnetic space?

The velocity of the magnetic axis through an electron equals, to some degree, the velocity of light otherwise the electron could not be maintained. This keeps the electron stationary, for the photo-electric field would not know if it was moving along the axis as light or whether the axis was moving down through the photo-electric field and around to maintain

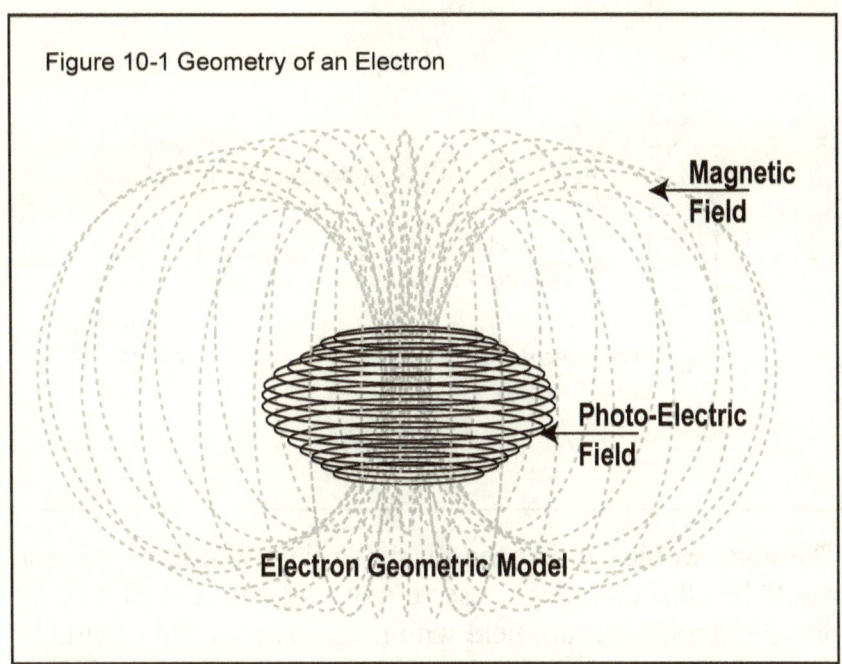

Figure 10-1 Geometry of an Electron

its particle nature. This is a better explanation as magnetic space replaces empty space. The only thing we would not know is how fast is the angular velocity of the photo-electric field? We might assume that this motion is the only constant motion in the universe. We can logically determine some aspects of the angular velocity of this photo-electric field within an electron by considering a relative magnetic density on the surface of the earth. Whatever this density happens to be, it will cause a single photon to spiral from an electron at 186,000 miles per second. I do not mean angular velocity—rather a linear velocity relative to the observer residing at rest in the earth's magnetic field. If the question arises in regards to the movement of the magnetic field from North to South, experiment would have to be done to determine the effect. The original Michelson-Morley experiment was done east to west from a distant star—never considering the Earth's magnetic field density.

Now **Figure 10-2** considers two concepts of light relative to the same magnetic density. The representations help convey the field theory of light in and outside of an electron.

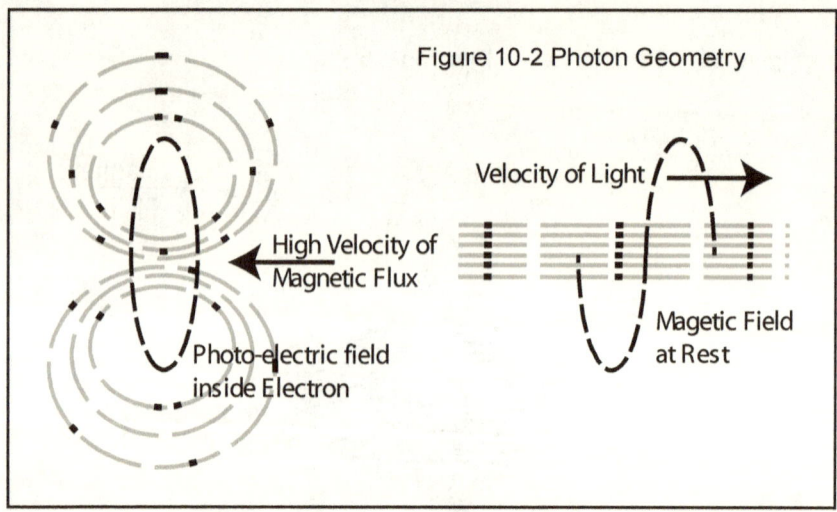

Figure 10-2 Photon Geometry

The logic would be that the magnetic field in circular motion would be 186,000 miles per second through the axis of a single photo-electric quantum field within an electron. This would be equal to the velocity of light outside the electron traveling in the same magnetic density at rest. Conservation of the photo-electric field's angular momentum relative to the magnetic field is conserved. If the flux density drops suddenly within the electron, do to the surrounding space, the result is that a certain amount of photo-electric quantum field would spiral off the axis and travel outside the electron? The velocity would be less near the confines of the electron and faster at a more distant point, because the relative density is not constant. Since photons come in discreet units, then it makes perfectly good sense to say that the photo-electric field is composed of discreet units or individual monopoles having momentum. The greater number of monopoles in the plain of propagation the larger will be the diameter of the photon. This provides for a greater degree of refraction.

In addition to the above, either an electron must carry its own magnetic properties or share with the atom. An electron is not apt to leave the surface of an element as easily as might be assumed. The greater magnetic density on the surface would attract the electron and provide a medium for static electricity.

The concept should illustrate that in order to keep the existence of a particle such as an electron; both the photo-electric and magnetic fields need to maintain a particular quantitative density. Just as the magnetic field changes dynamically the same can easily happen to the corresponding photo-electric field. The photo-electric field has a constant angular momentum, but the magnetic field follows a different rule of motion. This is the basic reason the two fields cannot be of one basic substance. Both fields vary in quantum density, but it can be easily assumed that an electron can absorb or expel various sizes of photo-electric quanta in order to explain the different sizes of electrons. There is no reason for baring this because physical bodies do absorb high frequency and emit low frequency, thus there is no reason to limit particles from doing the reverse. Such is the case in static electricity.

If we were to imagine an increase of the magnetic axis density of the circulating magnetic field, the velocity of the magnetic field through the axis might be less. Likewise if we were to increase the density or number of photo-electric fields, the magnetic flux density would need to increase. The argument to justify this conclusion comes from understanding the nature of a transformer. **Figure 10-3** will illustrate.

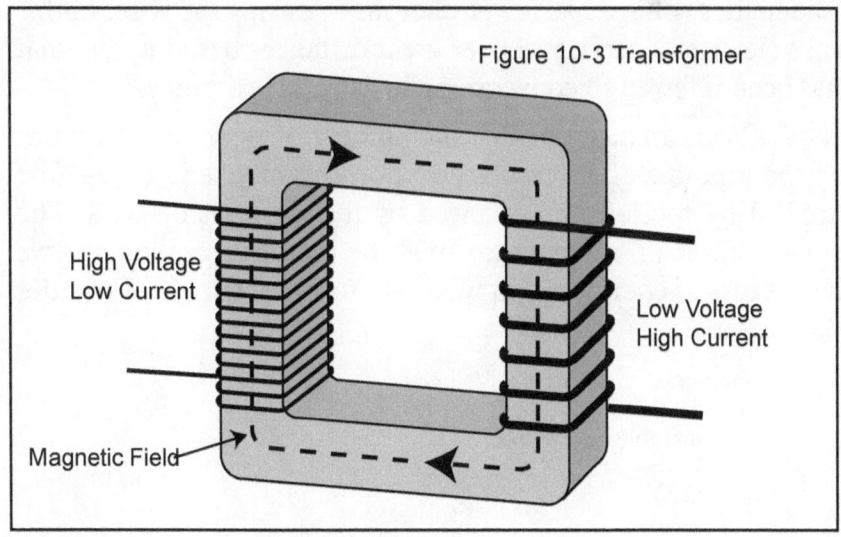

Figure 10-3 Transformer

For comparison we will liken each circular round of wire on the left to represent several photo-electric fields stacked as we would imagine in a large particle. We will also liken the current to these electric fields to be moving at near the velocity of light which would in turn move a conductive magnetic field through the circular iron core. Do not follow conventional knowledge and believe that the current in the wire is the movement of electrons. Consider electrical current as low frequency light—somewhat slower than in a vacuum due to the high density of the copper medium. Electricity does move close to the speed of light. As suggested earlier, the magnetic field moves within the core at a velocity similar to light depending upon its density. Note the number of wrappings on the left side as compared to the number on the right. What we have is a voltage drop and an increase in amps on the right side, but the velocity of the current would be the same in both sets of wires. Voltage defines an individual electromotive force but the amps define the total amount of current. We thus have a conservation of watts on both sides with less current but greater volts on the left and less voltage, but greater amps on the right. Think of the electrical current on the left as a composition of fewer but larger photons while on the right we have more photons but of a smaller size. I think further study would prove this to be true. The higher voltage has fewer photons to escape the wire. In this case electricity can travel over great distances based upon what has been referred to conventionally as "less resistance."

You can also think of the number of wrappings of wires as the amount of force placed upon the magnetic field—first applied by the left and received by the right set of coils. The conservation of momentum must be conserved otherwise we have a loss of energy and momentum from left to right. Consider the formulas:

Momentum = Mass x Velocity
and
Watts = Volts x Amps

Just as in momentum, watts are also conserved. Increasing mass and lowering velocity is like increasing the volts and lowering the amps. Both equations are conserved. I do not mean to equate the equations, but it does help to visualize that a higher voltage is a higher mass photon. A higher voltage would then require fewer photons and a smaller wire would work fine. A larger wire would be needed to carry many photons even if each is smaller. That is the essence of a transformer and why the wire is of different sizes. The great number of windings on the left is really equal to the fewer windings on the right in terms of the total momentum of the photons. It is this equal momentum on both sides that drives the magnetic field and passes energy from one side to the other. The reason I use momentum rather than just watts of power is to conceptually illustrate the transfer of energy due to an unchanged momentum. Whatever applies to momentum also applies to energy for both are conserved. The momentum of the high voltage current moves the magnetic field with the same momentum as it moves a lower voltage..

In order to do this with a fewer number of coils on the right, the magnetic field must move a greater number of photons of less voltage or mass. This is what is called a lower frequency. If less turns determine a lower voltage, momentum is conserved by moving a greater number of smaller photons than were used to generate the magnetic field. If photons are considered as mass, we can more easily understand the concept of momentum. The difference in traditional momentum of colliding bodies and the above is that the photo-electric field passes and receives momentum.

To understand the concept further, think of a spark of static light having the force of many thousands of volts, but the current equals that of a single photon. Keep in mind that a spark is light emitted by the electron of a visible frequency and the friction collected over time was adding to the electron many unobservable low frequency photons until it burst with one large photon because the electron could no longer hold the accumulated voltage relative to the magnetic field density surrounding the electron. Each surface electron can hold only so much voltage or only so much light can be crammed into it. The degree of this cramming is dependent upon the amount of the electron's own magnetic field it creates and that shared with atoms that create the needed density.

Look again at the transformer and think of each side as an electron. The left side has larger but fewer photons while the right side has smaller but many more photons. This is not the conventional way of expressing what happens, but it does explain the possibility that electrons can absorb one size of a photon and emit another.

Historically it was determined that when light strikes an electrical conductor it causes electrons to move away from their original positions. The observed phenomenon could only be explained by assuming that the light delivers energy in definite packets. Greater intensities of light (more photons) at one frequency (size of photons) can cause more electrons to move, but they will not move faster. In contrast, higher frequencies (larger photons) of light can cause electrons to move faster. *Therefore, intensity of light controls current density, but frequency of light defines the voltage or size of the photon.* These observations raised a contradiction when compared to classical sound and ocean waves, where only intensity was needed to predict the energy of the wave. In the case of light, frequency appeared to predict packets of energy. Something was needed to explain this phenomenon and to reconcile experiments that had shown light to have particle nature with experiments that had shown it to have a wave nature. Field theory as expressed in terms of particle density does just this. Things change because light has smaller break-apart units.

The magnetic field in the transformer is the carrier of energy. It does not matter what size or number of photons push the magnetic field or what photo size or quantity is pushed. It is the total momentum that is transferred. The right side of the transformer represents an increase in intensity (amps or more photons) with a drop in volts or light frequency (smaller photons).

A larger wire is required to carry the lower voltage because there are a greater number of photons. It is like many cars needing several lanes while a large truck needs only one. It is like a higher frequency or color on the left side being converted to a lower frequency of color (volts) with a greater current intensity (amps). I use color to accentuate the nature of light in electricity. Exactly what voltage belongs to a particular frequency has not been considered sufficiently. We must understand, however, that a volt is not a measure of energy. This is why momentum was used as a comparison for watts rather than energy. What is a measure of energy is what is referred to as an electron volt. A 100 watt light bulb burning for one hour is equal to 2.2 x 10^{24} electron volts or 2.2 Trillion electron volts—abbreviated as eV. The electron volt is the energy gained by an electron that moves across a positive voltage of one volt. For example 1.5 electron volts is the energy gained by an electron moving from a negative metal plate to a positive plate which are connected to the terminals of a common 1.5 volt "C" battery. Thus, visible light is composed of photons in the energy range of around 2 to 3 eV. It is the energy range of 1.8 to 3.1 eV which triggers the photo receptors in the eye. Lower energies (assumed to be longer wavelengths) are not detected by the human eye but can be detected by special infrared sensors. Higher energies such as x-rays (assumed to be shorter wavelengths) are detected by x-ray sensitive photographic film or again by special devices. How then does this help us in determining what 110 volts would be in frequency or color? It probably cannot because a volt is more like a measure of momentum while electron volts are a measure of kinetic energy.

More voltage (larger photons) is not more energy unless you also have more photons. Likewise, more photons do not increase energy unless you also increase the voltage of each. You must increase both intensity and/or the size of the photon to increase the energy. A higher voltage is a higher frequency or color. A higher voltage could just as well be a larger photon with a larger photo-electric field spiral that will refract more easily causing us to think in terms of a shorter wave length. It takes the net relationship of both volts (photon size) and amps (density of photons) over time to determine the energy.

> An electron volt is also the energy of an infrared photon with a wavelength of approximately 1240 nm. Similarly, 10 eV would correspond to ultraviolet of wavelength 124 nm, and so on.[7]

Higher frequency is more energy per photon, because it is more massive. But if you increase the voltage or frequency and reduce the number of photons proportionally in that the watts remain the same there is no net change in energy as there is no net change in momentum. In the above it is natural to assume that 10 eV has more energy than 1 eV because we are talking of either more small photons at a given moment in time or one larger photon, but the use of 10 eV would be equal to a wavelength of 124 nm (nanometers) and could only apply to a single larger photon with a wavelength of 124 nm or a frequency of 30 PHz (10^{15} Hertz). This is in the ultraviolet range. A photon having the energy of 1 eV would have a frequency of slightly less than 300 THz. This would suggest that to move the electron would take the emission or absorptions of a photon just below the visible range. Thus electrical current would not be able to move electrons—rather it will only slightly energize electrons and use them as a medium for passing photons from electron to electron. To think of electrons as current is incorrect.

Before we come to a point intended, keep in mind that physics does not measure the length of a photon as in wave length. It measures the refraction of the photon—the larger the photon the greater its refraction. The use of wavelength is only a mathematical convention and not a geometrical reality. The calculation is a left-over method use to measure sound and water waves. This is the reason for the expression "assumed wave length." As illustrated earlier, a large photon has a larger helix diameter with a corresponding greater number of photo-electric particles in the plane of propagation. This corresponds to a greater mass, a greater refraction and also a greater voltage. In other words visible light has a greater voltage than infrared light. The number of infrared photons is far more excessive than photons in the visible range. It is like a drop in voltage, but an increase in current. In the reverse, many infrared photons can pile up on an electron, but emission may be in the visible range. It is like many small photons accumulating to become a single large one. **Figure 10-4** is a geometrical representation of two photo-electric fields represented by a quantum number of photo-electric particles. The larger one would be a higher color temperature and also a higher voltage. It would also refract more when entering and leaving at an incident angle of a greater density of magnetic properties due to incident matter.

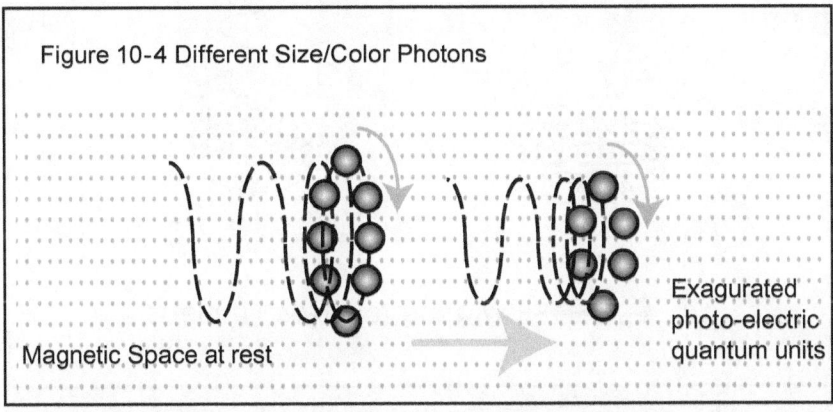

Figure 10-4 Different Size/Color Photons

Magnetic Space at rest

Exagurated photo-electric quantum units

We can now refer to the fact that light does move electrons in discrete units and these electrons carry the charge light gives them, but electrons do not move through a wire at close to the speed of light because the photons in electricity do not have enough energy to move electrons to that degree. If electrons were moving from place to place as conventional knowledge would suggest we would see visible light emanating from the wire. Electrons do not move freely, but are excited by light as it is absorbed and emitted. Charge is not a mystical bunch of plus signs or minus signs surrounding an object. Charge is the amount of the total photo-electric particles and the respective fields held in an electron by its own magnetic field. I use charge and not frequency because an electron is lifted through the use of many accumulated low voltage photons of electricity in order to provide the maximum intensity the electron can hold in its present shell or magnetic condition. It can become high enough that what is finely emitted is a very high volt photon that can be seen and at the same time allow the electron to move closer to an atom. Now the limit is determined by the magnetic field. **Figure 10-5** illustrates how the magnetic field plays a part.

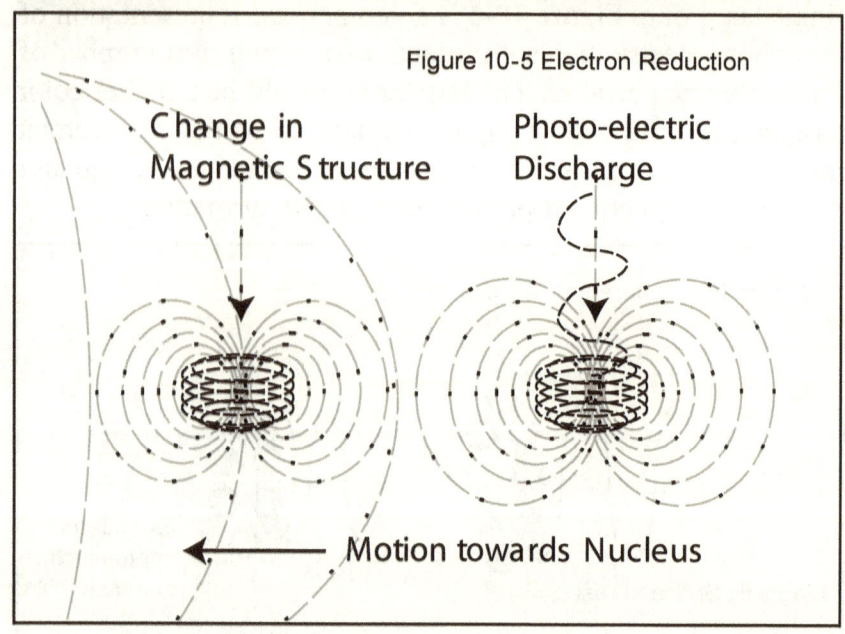

Figure 10-5 Electron Reduction

The concept is that for a free electron to exist it has to be able to generate its own magnetic field. If for any reason the magnetic density drops or is not sufficient to maintain the total voltage, the electron will lose an appropriate amount of photo-electric properties in the form of a spiral. This emission is in the opposite direction of the magnetic flux indicated by the down arrows. The velocity of the photon would be relative to the magnetic density in its path of travel. In this case the electron will move to a greater magnetic density in order to maintain equilibrium. Moving closer to a nucleus makes it possible for an electron to share a magnetic field generated by the nucleus rather than generate so much of its own. It must be understood that electrons do not fire photons as last in first out. The electron is more selective than that. It can absorb several small photons and emit one large or absorb one large and emit several small. It all depends upon the need to maintain magnetic equilibrium.

Action at a Distance | 202

The size of the photon is essentially the size of the voltage and the intensity is the number of photons emitted. In certain chemical reactions found in the two sides of a battery, a certain size voltage potential exists between the two poles. This means that only a specific volt photon can be released on one side and absorbed on the other. Such a voltage is so low compared to visible lite that a specific medium is required having electrons at a harmonic level to the electrons inside the battery. The electrons in the medium are aligned specifically to receive a kit in the ass so to speak in order to cough up the same energy or the same size photon received. It is the exact opposite of an iron filing remaining stationery but is aligned to the magnetic flux. The very difference between certain size photons traveling within a wire creating an electric field and the actual spiral field within the light is the difference found in referring to one as an electric field and the other as a photo-electric field. Thus the photo-electric field (number of monopoles in a single photon) and the electric field (a string of many photons), such as current in a wire, expresses different concepts of electric matter.

I suggest that the reason for the conventional negative charge is the polarity of the electron as compared to the polarity of the atomic nucleus. If they are not opposite in polarity the outer magnetic fields of the nucleus would not agree with the magnetic flux of the electron. Also the flux velocity of the electron's own magnetic field would be greater than the specific magnetic shell velocity of the nucleus, for magnetic fields are weaker or less dense the further distant from any axis. Since the velocity of light is an important aspect in understanding movement of the electron and the emission of light, we should understand that there are two aspects of the velocity of any photo-electric field in a single electron. One is the angular velocity and the other is the linear velocity when it spirals off. If all light of the same voltage has the same angular velocity and if we were to assume that a very long, low frequency, photo-electric spiral were able to be stretched out causing the spiral to move in a much longer spiral as one would stretch out a spring, we could suppose that the linear velocity is more equal to the angular velocity of light.

An increase in magnetic density, however, could create a slower linear velocity of light but not a slower angular velocity. This suggests that the magnetic axis is very close to the velocity of light in an electron and when it drops too abruptly or reduces the number of magnetic monopoles abruptly, a portion of a photo-electric field within the electron would spiral off and the electron would be drawn closer to the nucleus in order to find a pocket more suitable. Also a slower magnetic flux would need a greater magnetic density. These relationships are derived from the variable velocity of light in a variable magnetic density. These two concepts are what dissolve the Special Relativity Principle.

    In terms of the electron we can understand that the slower velocity of a magnetic field at a large distance from the nucleus would decrease in density as well as velocity but the larger the atom the greater this density and thus velocity at a given distance would surface. Understanding this helps in understanding why an electron will move closer to the nucleus yielding better conditions for a smaller electron or away if there are more photo-electric fields able to generate more of its own circulating magnetic field. At certain intervals the electron would have to give up light to fit in the confines of a smaller magnetic pocket or obtain more of its own magnetic field and share less with the nucleus in order to find a position further out. Finely, if the photo-electric field is smaller in circumference, its angular velocity would be at some point equal to the velocity of electricity and would move in a small copper wire or a large wire at a similar velocity, but the angular velocity could be less than the larger photo-electric field—meaning cycles per second. This only suggests that the real constant might only be the relationship between the two fields when one is denser than the other. All of these variables would determine at what point a certain size electron will feel most comfortable with a certain magnetic flux. In very dense areas within an atom, it is quite easy to begin to formulate the reason for electron shells without the electron falling into the nucleus. Charge has nothing to do with action at a distance as in gravity. It is simply a point of polarity and equilibrium depending upon the density of each field.

At higher levels the electron can be excited by the absorption of many low frequency photons until the larger excited charge is released due to a lack of magnetic flux. Again, the electron is not the measure of electricity. It is the 110 volt photon in massive quantities that forms electricity. Electrons come in many sizes. It is all due to the amount of photo-electric monopoles they possess. Disturb the magnetic field that holds an electron together and it will emit a certain number of photo-electric monopoles creating a specific frequency or voltage of light. Electrons can absorb small frequency over time and in one jolt emit one big photon. This would happen more to a valence electron than one deep inside an atom. Electrons deep inside are less likely to absorb photons in high frequency unless they are to change shells. The heavier the atom the more likely the electron in close proximity will have very little of its own magnetic field. It will become completely dependent upon the high magnetic density of the very heavy atom. Electrons are not pulled to the nucleus because of charge—rather they must give up in order for the electron to obtain an appropriately closer position. It is the magnetic field generated by the nucleus that locks the electron in its place and in the electron's own magnetic field that prevents it from getting too close. This situation is more magnified as we move up the element chart. Cesium for example has a size of 265 pm (pm=picometer where 1 pm=10^{-12} meter) yet Polonium is only 167 pm in size having an atomic weight greater than Cesium. Polonium has 29 more electrons and 29 more protons than Cesium but is still about a third smaller in size. The conventional argument tries to explain this by saying that the increase in protons attracts the electrons closer. If such was the case, why do not the electrons just fall into the nucleus? The electrons are pulled by the stronger magnetic fields generated by the nucleus, but this is still a poor description. Actually the magnetic flux density is greater in the heavier elements allowing for the closer electrons to give up their magnetic fields and rely almost entirely upon the flux of the atom. With fewer magnetic fields on the part of the electron there is less to oppose or less magnetic field going the wrong way. They can now fit tighter without a loss of light and thus reduce the size of the atom. As far as the element chart is concerned, when you move to the left and start the next element with one electron more there is a creation of a new outer shell and thus the atom increases in size only to be shrunk again as the element increases in atomic weight to the right.

Whatever is measured more accurately in science becomes better able to be explained by field theory. Field theory proves to exist more easily than before, yet it is completely neglected in text books. There is one discovery, however, that seems to confuse field theory in some respects and agree with it in another. It was Otto Stern and Walther Gerlach in 1922 that devised an experiment to determine whether particles had any intrinsic angular momentum. It is the nature of spin when it comes to the electron. In a classical system, such as the earth orbiting the sun, the earth has angular momentum from both its revolution around the sun and the rotation about its axis. If the electron is treated like a classical solid-like dipole with two halves of charge spinning quickly, it will begin to precess in a magnetic field, because of the torque that the magnetic field exerts on the dipole. Despite this usual description, theoretical studies have shown that the spin possessed by these particles cannot be explained by postulating that they are made up of even smaller particles rotating about a common center of mass. It is thus said that these elementary particles are true point particles rather than a composition of interacting fields as suggested above. The spin that these point particles carry is concluded to be an intrinsic physical property, akin to a particle's electric charge and mass. What is said causes more confusion than admitting the existence of a field of monopoles rather than a assuming a mathematical point. I think that classical physics cannot be applied to the electron as it can to solid-like particles. As soon as you do so you end up giving the electron magical powers or, as said, intrinsic physical property that cannot be explained. This is the nature of physics. For everything they cannot explain they attribute action and reaction and simply give a mathematical basis with a strange term such as charge. We are left with something that is not really existent but is only a convention to explain the action.

The usual comparison with classical dynamics in order to explain things and then turn against such methods is the nature of physics today. The reason is that the theorist falls prey to the equations that treat mass as a point-like particle rather than a field that occupies space. The theorist is pulled between the equation and some classical attempt to explain. You can search the internet for hours on end with many honest questions that are really never answered. The questioner is finely put in his place by someone retorting the equation process. As far as an electron having spin in two directions is actually dismissed in the classical sense—rather it has a symbolic up and down spin because when the electron passes a vertical magnetic field it either turns up or it turns down. Trying to explain this in classical terms is virtually impossible because an electron is not a solid particle. It is also not a point in space as the equation needs in order to make a calculation as to its momentum. Those falling prey to the equation end up as mathematical robots predicting every event without the slightest bit of geometrical understanding.

The field theory as has thus been expressed can explain electron spin both geometrically and intrinsically. It is generally determined that the experiment removes any extraneous fields in order to determine a net spin. The problem with this is that space itself is magnetic. I do not say that it is cable of applying a force—rather the electron must of necessity align itself to magnetic space if moving at high speeds through it. By simply firing electrons into space at rest must of necessity cause the electron to either align its north pole in the direction of travel or its south pole in the direction thereof. This is due to the passing magnetic monopoles at rest, but the electron thinks them to be raining upon it. **Figure 10-6** will explain the Stern-Gerlach experiment in terms of field theory. The polarity of the electron would naturally align with either the internal flux axis or the external flux of the electron. Between these states would of necessity force the electron to adjust its axis.

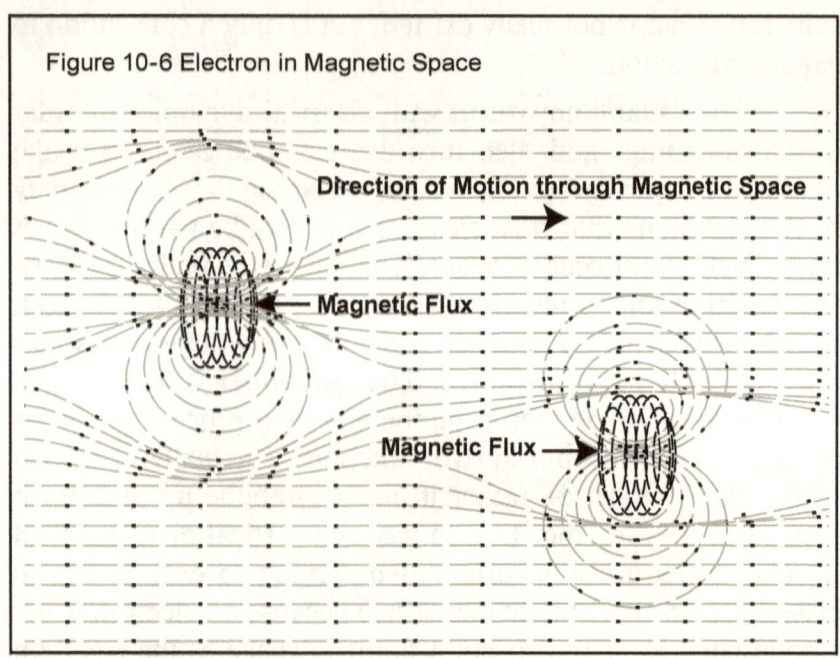

Figure 10-6 Electron in Magnetic Space

Contrary motions of the magnetic field are avoided. In the two examples the magnetic field in space splits according to the oncoming electron as the electron aligns itself to one of these two positions. Electrons come in random orientations, but high velocity through magnetic space sets up the electron to align itself into these two directions. This process would be referred to as the intrinsic nature of the electron to align the poles in the direction of travel. Now when either of these two electrons passes a magnetic field perpendicular to the direction of travel, we have a situation where the leading pole of the electron is going to experience a slight pull up or down depending upon the intrinsic position of the electron. This slight pull will alter the direction. Once the electron leaves the magnetic field its new direction has been established and will remain in that direction until a collision is imposed. This is not the typical refraction of an electron. It is the result of intrinsic polarization. The use of spin up and down is really not a good explanation—rather the spin is clockwise or counterclockwise in the direction of travel. This is the geometrical explanation. Electrons only have one spin as in the right or left hand rule but not both. If you change

Action at a Distance | 208

the spin you simply turn the electron around. You polarize it to other particles. This is a far better depiction of spin.

There has been no attempt to change the mathematics or the quantum effect. The attempt here is only to give a geometrical analysis for which modern physics is so inept to do. Field theory and magnetic density expresses a far better conceptual explanation than using relativity and the uncertainty principle to find the probability of an electron. The very same conclusion can be drawn in regards to protons, neutrons and quarks as having this net polarity. Some particles such as baryons do not behave the same, thus suggesting more complex pole configurations. Photons and other gluons have no intrinsic polarity suggesting there measured existence is motion around a magnetic field at rest in the same manner as light. Understand that the nature of polarization for light is a very different concept to be discussed in a later chapter. Light has only one spin with a virtual polarity in the direction of travel, but since the magnetic field is at rest there is no tendency to change directions other than a certain degree of refraction if one side of the photon has a greater magnetic density than the other. The following is something that Newton said in this regard to a prism:

> Every ray of light has therefore two opposite sides... And since the crystal by this disposition or virtue does not act upon the rays except when one of their sides of unusual refraction looks toward that coast, this argues a virtue or disposition in those sides of the rays which answers to and sympathizes with that virtue or disposition of the crystal, as the poles of two magnets answer to one another...
>
> Newton 1717 [8]

The two ends of the photon are polarized in the sense of magnetic space. One end is receiving the magnetic axis and the other end is expelling it. Newton was not really saying this in terms of forward and afterward. It was the diameter from the two sides (Newton's attempt) having different reactions to the different densities of magnetic space.

In determining the nature of charge as intrinsic to the helical nature of light is simply increasing the size of electrons with a polarity opposite that of the nucleus. This should illustrate that the electric field does not emanate from so-called charged particles. Field emanation belongs only to the magnetic field as far as action at a distance is concerned, unless the physicist wants to conjure up a third unknown field. Light emanates, but this represents actual travel over a specified distance of a piece of the electron. The absorbing and emitting particles are not connected in any way but to the magnetic properties of a larger particle. To think of charge as fields that are not magnetic is to make up something to fit the behavior rather than use concepts as Faraday had attempted to demonstrate. An electron has charge because it is pumped up with many low frequency photons. The photo-electric field itself is light and the electric field or current is the number of photons absorbed or emitted by the electrons that hold this charge.

Just as light comes in certain sizes, so does the electron, which has many quantum levels or a large number of electric monopoles. Also, just as the many photo-electric particles create a photo-electric field within the photons, the photons themselves create the electric field through a wire by leap-frogging from electron to electron. In addition to this the electrons create an electron field about the nucleus. The many electrons sharing the external magnetic shell generated by the nucleus form the electron field while the photo-electric field is within each electron. The electron field was developed from a mathematical concept of which the field of electrons would orbit around the axis of the atom. This planetary model, suggested by the Danish physicist Niels Bohr (1865-1962), suggested a cosmic-like system. One might easily suppose a pervading magnetic field circulating within the center axis of this atomic structure. Every photon leaving the nucleus creates a proton in neutron decay and defines the electromagnetic interaction deep inside the atom. This interaction includes the magnetic field packed tightly and strangely polarized to produce a nucleus. Either a net circulating magnetic field moves with a gross particle or light spirals through a magnetic field at rest upon leaving a particle.

Action at a Distance | **210**

In this the magnetic field plays its part in the process of both light and matter.

Since the Bohr model there have been changes due to electron spin and certain bonding characteristics. With this in mind, it is important to cover the reason why electricity needs a medium of electrons and why certain electrons in specific elements do not contribute to this conduction. If it is concluded that the greater magnetic density resides closer to the nucleus and increases as the atomic weight increases and the size of many atoms decrease in size as the atomic weight increases, we can generate a better understanding about element conductivity. Traditionally it is considered that unpaired electrons contribute to conductivity. It is thought that paired electrons offer opposite spins and thus would cancel each other out. This would inhibit conductivity as well as magnetism. As illustrated, if electrons were up and down or were polarized they would not agree with a magnetic field of the nucleus. Paired electrons may be considered essential to the bonding of two atoms but not deep inside the atom. Electron pairs would be nothing more than two electrons having a specific orbit density—as if they were orbiting the atom with the same polarization in the same orbit but on opposite sides. It makes sense to have two electrons to balance the centripetal force. Experience with paired electrons in bonding does not assume that every other electron is polarized to another. This was perhaps due to a misunderstanding of the Stern Gerlach experiment. What should be concluded is that polarized electrons could create a bond between two atoms. This could be explained in the discussion of conductivity. **Figure 10-7** will illustrate this approach.

The chart of elements is different than most charts due to the inclusion of the inner transition elements usually in a separate chart below. The chart indicates that electrons are not added on top of each other in an orderly fashion. It depends upon the magnetic density which varies at distances as each element shrinks up the scale until a new ground orbit is added. Iron, aluminum and silicon are illustrated based upon their

Figure 10-7 Chart of Elements

Shell	Ground Orbit		Transition Elements										
1	1												
2	3	4											
3	11	12											
4	19	20	21										
5	37	38	39										
6	55	56	57	58	59	60	61	62	63	64	65	66	6
7	87	88	89	90	91	92	93	94	95	96	97	98	9
	1st Electron in New Shell	2nd Electron in New Shell	Inner transition elements are the la electrons placed below ground orbi Shell 6 electrons are really in shell These elements are usually placed outsid the element chart.										

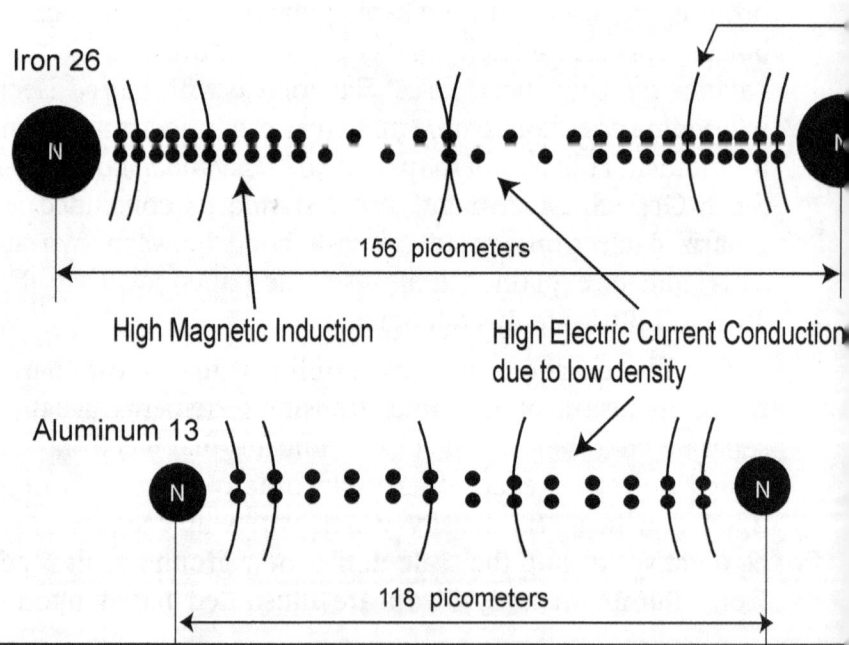

Iron 26

156 picometers

High Magnetic Induction

High Electric Current Conduction due to low density

Aluminum 13

118 picometers

torns between ground orbits

ium is really the second electron
ed in the ground orbit.

										Orbits Above Ground					
															2
										5	6	7	8	9	10
										13	14	15	16	17	18
		22	23	24	25	26	27	28	29	30	31	32	33	34	35	36
		40	41	42	43	44	45	46	47	48	49	50	51	52	53	54
70	71	72	73	74	75	76	77	78	79	80	81	82	83	84	85	86
102	103	104	105	106	107	108	109	110	111	112	113	114	115	116		

These transition elements are placed in the central orbits between ground orbits.
Shell 4 electrons are really in shell 3.

These electrons are placed just above the ground orbit of the shell they reside.

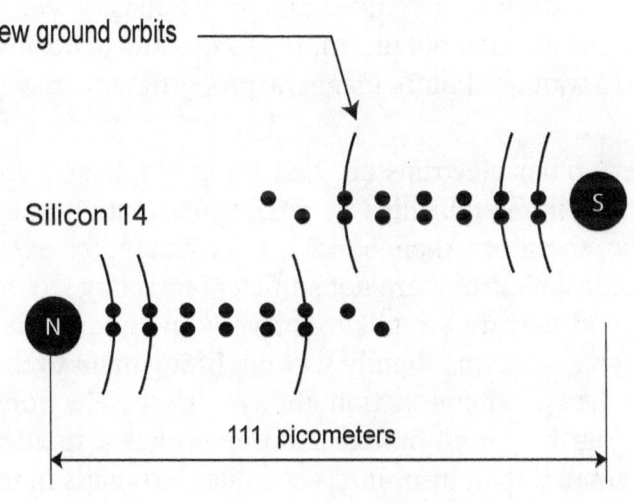

New ground orbits

Silicon 14

111 picometers

Lower Magnetic Induction and Electric Current conduction could be due to polarity more than density.

radius from nucleus to nucleus and also their traditional electron configuration. For simplicity all electrons are placed in one equatorial line for comparison. Normally electrons are spherically all over the place. We have to understand that electrons close to the nucleus have little of their own magnetic fields and can sit closer to each other. The examples depict current theory, but it should be expected that the distance between electrons would increase the further from the nucleus in order to accommodate more magnetic properties for the electrons. The electrons further out have more room and more of their own circulating magnetic fields and can be energized more easily and thus act as a conduit for electricity—not necessarily in a straight line. The extra spacing indicated above is due more to the current theory that electrons are not paired with one orbit at a time, but several orbits are filled with one electron before they are paired below that level. Two electrons are placed in the ground state for the next shell before the lower shell is often completed. Higher up the scale, shell four electrons are really filled in shell three. This is the reason for the step in the chart. This delayed placement indicates that the density closer to the nucleus is not sufficient for the electron to exist with a sufficient magnetic density which will come when the element obtains more protons and neutrons. Electrons have to wait until more magnetic properties are drawn into place.

The very outer electrons are tied up in bonds and are also limited in their relationships in passing current. If they were able to be energized their bonds might change or even break. The photons in current are not sufficient in voltage to do this. Keep in mind that we are talking of very low frequencies that could energize electrons slightly without lifting them to the next shell. Iron has good conduction ability with the electrons just below the fourth ground orbital. Aluminum has a smaller nucleus and is smaller than iron. It has a similar looseness in its outer shell. Silicon, however, is still smaller in size yet is packed with only one more electron over aluminum. In the case of iron and aluminum only the outer orbits overlap. This would be in agreement with the sharing of electrons. Silicon is not a good conductor eccept at very high voltage. This is what makes it a good switch in transistors. The question is, how can silicon compact to such a small radius with only one more electron than aluminum? Field theory cannot explain this overlap unless the bond differs. I would suggest that every other atom is polarized. This would explain the neutrality in magnetic induction and electric conductivity. The example does not show the boding of polarized atoms. It is the overlap that cannot be explained and therefore I have left the two atoms as shown without bonding. Some complex arrangement would be required to explain the several orbit overlap.

Using the above facts, why does aluminum have a low induction for magnetism, yet a strong structure for electrical conductivity? Magnetic induction differs from the conductivity of light or current. Even though both fields are moving linearly in opposite directions in order to maintain the electromagnetic relationship illustrated thus far, current is inhibited among high density electrons while the magnetic field is more apt to move through this high electric density. Now some elements have a higher density, but their molecular arrangement may have contrary polarization and thus inhibit magnetic flux through certain molecules. This is where the polarization and the type of bond create the difference.

It might be appropriate to point out that in our transformer the momentum of a direct current does no altogether pass from one voltage to another. It might provide a nudge when the current is turned on, but will dissipate quickly. And even though the direct current continues on one side the receiving side does not endure. It takes an alternating current to produce a constant change in voltage. Why would this be? We must understand that the electric field is specifically confined in density and direction while the magnetic field is not. The magnetic field will dissipate very quickly by spreading its momentum in all directions by actually leaving the coil. The most interesting point about the magnetic field is that at close distances from one magnetic monopole to the other the reaction is specific in momentum exchange, but this momentum dissipates in all directions unless there is an alternating current. With an alternating current the momentum exchange within magnetic space is limited to very short distances before a reversal of momentum is instigated with equal energy. Before the magnetic field has a chance to dissipate it is forced to reverse. It helps to understand that the magnetic field in matter varies in direction and the momentum dissipates toward a greater electric density unless there is a change of direction. The magnetic field dissipates to the magnetic flux which in turn produces the nature of gravity. A constant reversal of direction creates a directional wave that does not dissipate. This will be explained in the next chapter. This explains why the magnetic field seeks more flux density in matter than when the electrons are spaced further apart. Outer electrons can have opposing magnetic fields and be of larger size because there is room to generate more of their own magnetic fields, thus absorb more charge but inhibit magnetic induction.

The majority of the electrons in iron have this tight characteristic which seems opposite the characteristic needed for each electron to absorb and emit light. It is the uncompressed electrons that provide the electrical conduction and the compressed electrons, including nuclear particles that provide magnetic induction when the atoms are aligned appropriately. Improperly aligned particles can inhibit the flow of magnetism. Electric conduction (absorption and emission of light) should not be inhibited as long as there are some loose electrons. Either magnetic induction or electric conduction would not have to move in a perfectly straight line. Keep in mind that the magnetic induction moves through each properly aligned axis and electric induction moves from electron to electron surrounding each axis. Some atoms may all be aligned similarly and others may even be able to adjust slightly. It may be that iron is bound differently than many other elements and thus provides the ability for conduction.

It should be made clear that elements are randomly bound and it is the net polarization that determines induction and conductivity. Bonds could be at the atomic equator or shifted in latitude. They could also be polarized bonds and each layer of elements could be bound differently than the first layer. Perhaps individual molecules can rotate slightly. Field theory becomes very complex and is the reason that it is conceptually avoided. It is easier to depend on equations than truly understand the geometry. The only problem with this process is that we define our terms to perform magically rather than conceptually.

The attempt has been to provide an electromagnetic explanation for electricity and electron levels. The electromagnetic relationship can explain all aspects of the elements and their relationships. The only item not clarified is why smaller low frequency photons perform better in a medium while larger high frequency photons cannot even penetrate the same. The spark from a wire's surface is a free electron having excess charge that eventually radiates from the surface of a medium. It is in the visible range. The electron receives so much low frequency current that it finely bursts. Electrons deep inside the wire do not do this. They have limits and any excess current is radiated off in many small photons of heat. Before that point every electron capable of swelling is like a sponge. It consumes as many low frequency photons that it can hold until every electron possible is at its optimum charge. Turn on a switch and the electrons in the uncharged wire absorb the thrust of photons coming in from the positive end. If a current of 110 volt photons pass through a tungsten wire designed to radiate by the passing of current, it is like certain electrons absorb the 110 volt photons to the point of radiation of visible photons. Certain electrons in tungsten absorb at low frequency and emit at the visible range. This happens over and over until the current is shut off. In the case of a motor it is the motion or alternating current of the photons through a circulating wire that sets up a magnetic field that turns the motor. As long as the photons flow they will do their work. This means the current must be grounded to something that will absorb the current or return to a charging system or battery.

It is interesting to understand that if a charging system continues with little or no ability of the electrons to absorb the current, the alternator in the case of a car will burn up. The alternator, for example, has an internal switching mechanism that when the voltage reaches its potential, it will shut off the generating current. Major power sources have to start up and stop generators consistent with the demand for energy otherwise energy is wasted in heat. Heat is really none other than small infrared photons leaving the area of overstuffed electrons. Heat is also conductive and follows the medium more than not. Lower frequency of light prefers a medium, but if the medium cannot receive the high frequency, light is reflected or refracted within the material. Low frequency light is more susceptible to absorption because the spiral helix is small compared to the space between electrons. This explains the inability for outside air to conduct heat as well as dense metals. Lower frequencies do better in the proper medium.

The next area of discussion as to field geometry is the nature of bonding. In **Figure 10-8** I give a concept of a single bond using two electrons

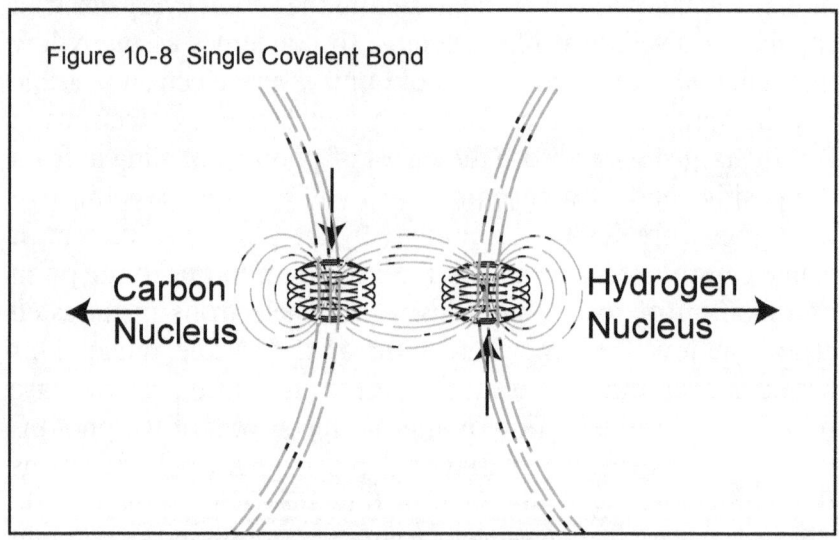

Figure 10-8 Single Covalent Bond

The illustration is conjecture due to the need of first suggesting a stationery possibility using the two elements as an example. The actual atoms, however, are slightly apart and are also polarized to each other. Experiment does not consider

Action at a Distance | 218

polarization because the magnetic field is never considered. A single bond does measure a greater distance between nuclei than a double bond. In other words the two atoms would normally have magnetic fields not in agreement if it was not for the two electrons to bring them together. If covalent bonding does not necessarily require the two atoms then one electron would hold two non-polarized atoms a bit closer but the magnetic density would not be as great. This would seem to be contrary to convention. The arrows indicate magnetic flow. By this it is meant that one electron binds by incorporating the magnetic field of the other electron. The electrons could become larger due to the added magnetic density. To do so would be the absorption of the appropriate amount of added light—usually referred to as energy that creates the bond. In this case the two atoms do not share electrons. If there is a second electron in the same orbital it is most likely used to bond with another atom in the same manner. Bonding either absorbs a certain frequency or emits it. Once this is determined it is easy to explain using magnetic density or strength needed. In the above example light would be absorbed due to an increase in size of the electrons. The opposite would be an emission of light if the increase in magnetic density is sufficient to use smaller electrons—especially if more electrons are involved. The concept of sharing an electron could happen if an atom in need pulled a valance electron from another atom, which allowed one orbit and returned it to the donor element. This exchange would create a figure eight arrangement. The distance of the atoms would be greater between nuclei than the bond given above. If an atom is short an electron, light could easily be absorbed to create one bonding electron. It all depends upon what is actually observed. As illustrated so many times, observation should agree with reason as the verb agrees with the subject. It should not predict an assumption lacking consistent good sense. Coming up with new magical forces under special circumstances does not agree with reason. New names are often invented just to explain misunderstood strangeness.

    In reflection upon the bond between silicon atoms and their excessive overlap until they complete their compressed size, it becomes a quandary not because one cannot speculate a field pattern, but because there is no experiment or theory that can pave the correct way. The precise type of bond would have to be considered rationally. The double bond may explain some of the contraction as the elements move from left to right. We know that the added proton, neutron and electron can increase density and thus contract the atom, but can we not also assume that when two atoms come together having the same polarization this would increase the magnetic density between them and thus allow a tighter fit using four or six valance electrons. Not only the number of elements could contract the atom, but the greater number of electrons could also bring the two atoms closer. **Figure 10-9** illustrates a remote possibility.

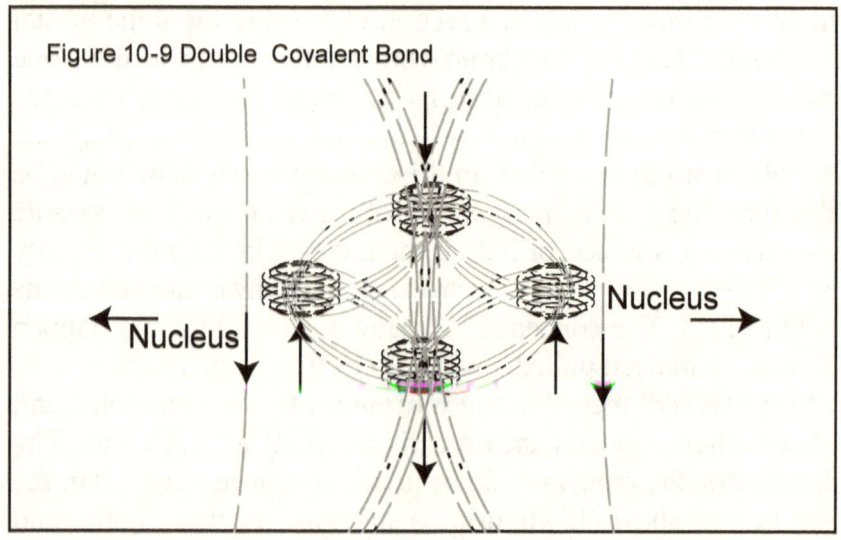

Figure 10-9 Double Covalent Bond

It seems that we are dealing with a greater amount of electric matter compacted in such a way that we must first question how that compaction takes place. I have suggested that the space between electrons condenses due to their ability to use less of their own circulating magnetic field and rely more upon the higher density generated by the combination of two non-polarized atoms.

The examples do not end here. You could have two very large electrons stacked above each other if you do not assume four electrons are needed. The two might do the trick. You could even speculate two electrons or even four circulate in a figure eight around the two polarized atoms. The concepts depend upon if the electrons are being shared or not and whether the atoms are polarized to each other. The examples could be many, both simple and complex. The more electrons involved and the greater the overlap the easier it is to understand the radial reduction between atom. Simple is always preferred, but experiment might demand a more complex solution. Relying on the understandable is preferred because you can too easily assume some magical force is not appropriate to reason.

Here are some examples that have been concluded: A single bond involves the sharing of two electrons as is the case of water. Between the hydrogen and oxygen each bond is a single bond and contains two electrons. In a double bond we have the sharing of four electrons as in alkenes or between oxygen and oxygen in O_2. In a triple bond six electrons are used in the nitrogen gas (N_2) between N and N.

In the above examples hydrogen is in the first ground state and the oxygen is in the second ground state at an almost complete condensed condition before a new ground state comes. Are the two elements polarized? If oxygen has two single electrons in the outer two orbits and hydrogen has only one, this means that two hydrogen electrons (one from two individual atoms) to a single oxygen electron (one atom) are used. There is no sharing in terms of one atom being without an electron in its pre-bound state. This would suggest polarization of hydrogen and oxygen to each other. In a double bond the examples give two oxygen or O_2. Are the two atoms polarized? Both oxygen atoms would have two electrons in the outer orbit. This makes for four electrons in a double bond. If the double bond illustrated above is the answer then the two atoms are not polarized. The electrons could then circulate both atoms in a figure eight if the graphic example is incorrect. This would mean that an oxygen molecule has no more than two atoms unless one atom has only one outer electron and shares it with oxygen having two. The extra electron of the one could then be used to bind another oxygen atom. Sharing and polarization need to be explained more fully in bonding. Too much is assumed.

In the last example each nitrogen atom has three single electrons in three outer orbitals. Again, are the two atoms of nitrogen polarized? The same argument would follow that of oxygen, but the bond would be stronger, yet it is difficult to speculate on their geometry if the electron remains stationery. This suggests that a figure eight might be the more appropriate, but we have to conclude that some electrons could be used if many atoms can combine. Since we are talking about a gas it might be that the number of atoms in a molecule is limited unless compressed.

One thing that has not been emphasized is that atoms drastically increase in size once a new ground orbit has been established. This is when an element starts on the left side of the next row on the element chart. This suggests a pattern in atomic magnetic field structure that is not an even or gradual magnetic density. It is as if the magnetic field obeys certain jumps due to the patterns of the nucleus. This might be a start in understanding the arrangement of protons and neutrons or even some special combining. Perhaps when a new shell is created the arrangement of the nucleus also changes. Also, in order for the atom to generate a magnetic field just to work its effect upon electrons it must generate its own magnetic field. This indicates an orbiting motion in the same direction as the electrons. It is either this or all protons have the same polarization that is opposite to that of all the electrons. If all of the protons were in the center and the neutrons surrounding, it would facilitate a tighter atom. Maybe the protons are random and their alignments only indicate their effect on the magnetic flux and like electrons in gravity the vertical alignments are neutral. This concept will be explained in the close of this section.

I am of the opinion that individual neutrons and protons may not exist together as a bunch of marbles. It is the whole nucleus that has a certain charge or polarity opposite of the electrons. The energy and the mass is quantized in each nucleus, but just because a neutron flies out does not mean that it existed exactly internally as it does externally any more than a photon existed internally as it did externally. All things break down to the smallest unit. If an electron can increase in size, there is no reason not to suppose that they can combine into even larger particles if the magnetic density is sufficient. It is either this or at a very large state they share the same circulating magnetic field. In either case, this is nuclear compaction.

Neutrons, for example, are unstable outside of the nucleus. If you put a neutron in free space and wait about 700 seconds, it will split and out comes a proton, an electron, and a nearly undetectable neutrino. What held these units together was the higher magnetic density within the nucleus. If a particle can exist outside it means that it is structurally a single unit of photo-electric and magnetic properties. If it is divisible such as three quarks to a proton or neutron, we must ask if a quark can exist alone. So far they cannot. This indicates that a quark is a relation more than a particle. A relation could be two electrons bound by a third or one electron being bound to a proton by a neutrino within a neutron. If a neutron has enough energy to leave a nucleus would it not already have a certain degree of angular velocity? You can split an atom, but normal decay should indicate that a nucleus can regularly whip a particle from an internal orbit. This would be because the magnetic density is not sufficient to maintain the atom that is probably on the surface of a very heavy element. Once it might have been created under great pressure or a high increase in magnetic compression, but once it comes to the surface (even inside a rock) decay happens naturally.

Further conjecture is difficult without more information that is consistent. Whoever has the taste for graphic computers, equations and field theory, might just change the direction of

The Einstein Illusion

physics forever. At this point I leave it into the hands of others keeping in mind the possible whirlwind that is happening inside of every atom.

I started with particles and did a simple trek to the nucleus which has various illusive qualities. Now for the largest of forces we call gravity. This too is nothing but understanding magnetic density. I refer to it as a graduated density unlike a step density in the structure of an atom.

Under a broad generalization, gravity can be defined as the earth's magnetic field reacting to every photo-electric field within each particle approximating a perpendicular orientation to the earth's magnetic field. Even though a particle total accumulation of electric properties vary in direction, every field reacts a bit with the earth's magnetic field as it passes through the particle's magnetic axis, and this does not affect a gravitational pull. In other words, the net flux or motion of the earth's magnetic field has little to do with attraction. It is the local change in magnetic density from one microcosmic elevation point to another that determines the cause of gravity. Add to this the common understanding of acceleration and the particle falls. This is the same as what causes light to bend as illustrated in earlier chapters. Note **Figure 10-10** for an analog of gravitation.

The force downward will only happen while the electric field within mass is polarized—the magnetic axis is parallel to the earths' magnetic field (an exaggerated individual field combination inside matter). Even though individual electrons and other particles change orientation constantly, the random arrangement of elementary particles implies that, on the average, approximately half of gross matter would be polarized perpendicular to the earth's magnetic field while the other half would lie parallel. Thus one half is accelerated to the earth and the other half resists motion as inert particle fields. Fields are random, but it is the net one half that determines curvature in gravity.

Action at a Distance | 224

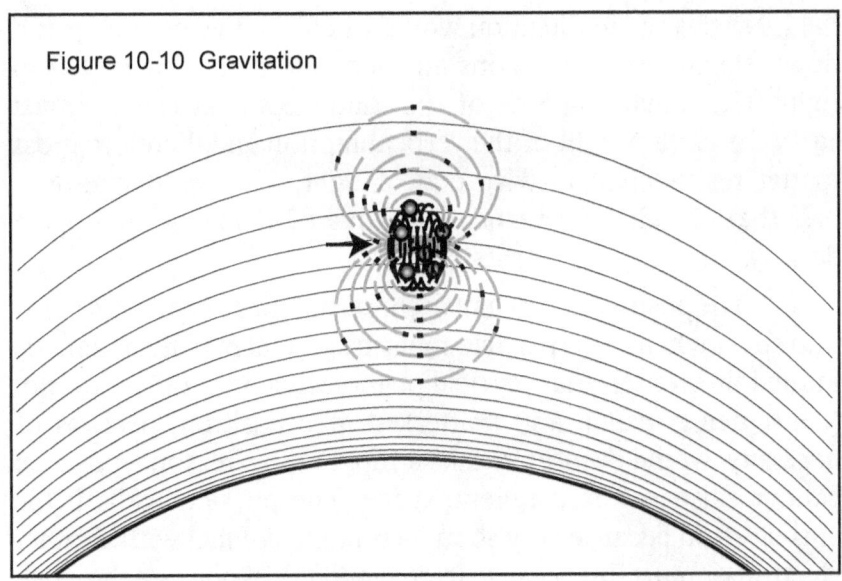

Figure 10-10 Gravitation

The magnetic density from the earth's surface to distant space follows not only the inverse square law, but also it explains why light curves twice that of ponderable matter. Keep in mind that this is not necessarily flux density or speed of the magnetic axis. This would be a net magnetic monopole density regardless of motion. However, if the motion of a magnetic field is in agreement with or opposite the particles' magnetic field it should have some Doppler Effect. It is important to think in terms of inverse square law particle density rather than flux density. The first governs gravity and the latter governs bonding and particle creation. Flux density would only balance out other internal particles having a magnetic field in disagreement. The reason that the graviton may never be discovered can be due to the idea that gravity emerges as an infinitesimal vector product of circulating electric fields residing in an inverse magnetic monopole density generated by the a larger mass such as the earth. The reason that light bends twice that of ponderable matter glancing past the sun is because light carries a pure horizontal relationship to the direction of the sun's magnetic field or surface. Inside the atom, a net 25% of electric matter spirals upwards, a net 25% downward and only a net 50% move horizontally with the earth's magnetic field. No change in direction would be effected when the up and down electric axis maintains an even field density from left to right. The remaining 50% of spiraling electric properties must carry the extra weight of the vertical motion and therefore gross matter resists change twice that of light. Thus matter refracts half that of light in a vertically graduated magnetic monopole density.

The positive and negative characteristics of particles come into effect when a free particle spins in agreement to the magnetic induction through that particle or molecule and others do not (conveniently called anti matter). A free positron, perhaps of necessity, would eventually take a flip. For this reason a positron cannot exist in free magnetic space. The physicist calls it the anti-electron because very soon it comes in contact with another electron or must flip because its magnetic field opposes the earth or testing equipment's surrounding magnetic flux. The angular momentum of the two coming in contact is contrary enough to distribute the two opposing electric fields into three spiraling photons that 'shoot off toward the apices of an equilateral triangle.[9] The origin of this positron comes from the breaking up a neutral particle in order to produce a positron and an electron. The positron and the electron do not easily recombine into a neutral particle. This would be due to the new parametrical conditions of the respecting magnetic fields. The magnetic conditions are now spherical instead of a possible original shared circular ring. The net result conserves what is referred to as electric charge and also the angular momentum. Physicists call the destruction of a neutral particle mass annihilation. Can it be true annihilation when the mass angular momentum of the two electric fields converts to the mass linear momentum of the three spiraling photons? Logically it can be said that if the positron would flip over, it would become an electron. But under ordinary conditions on earth, a positron "cannot move for more than a millionth of a second or so before it collides with an electron."[10] Consider that on the equator an electron exists free from an atom. With its magnetic axis pointed north and south, agreeing with the earth's magnetic flux, take it immediately to the center of the earth and observe that the electron will become a positron. It will probably clash with an electron before it has a chance to take a flip. This is facetious, but illustrates a possible explanation why we do not observe positrons. We forget the flux of the earth's magnetic field.

Confidence strengthens in the direction of a unified field theory when the magnetic field plays a part in all reactions—everything from atom to gravitation. Gravitation certainly pervades outer space as well as beneath the earth, and if any object carries simple particles such as electrons and photons, a reaction, however so slight, must exist. Time must eventually reveal that to unify the forces of nature, the scientist someday will see that behind the relationship of the electric and magnetic monopole fields of inertia, we find hidden the common denominator that will resolve the distinction between the two fields. Unity will only be reached when this right angle relationship or distinction in design can be understood. Therefore, the only mysterious cause and effect relationship thus far comes from the relationship between the magnetic field and the electric field respectively. Even this can be understood not as a right angle event, but as an angular effect. Consider the spiral current in a wire. It is going to have an angular expelling effect upon the exiting magnetic flux inside the spiral which would begin to spiral in like agreement to the photo-electric field and as it begins to spread out after exit, it follows a larger and larger exhaust spiral until it moves almost at right angles.

At least all of the paradoxes of relativity are removed by the very definition and existence of magnetic and electric monopoles.

A Particle of Intelligence

This section will continue in hopes of understanding the more elusive particles. Whether I am wrong or right, I would like to explain that I do not believe that intelligence is a bunch of protons and electrons or anything to do with electric matter. Electric matter is entropic while intelligence is anti-entropic. Intelligence evolves from a lesser order to a greater order. Electric matter dissipates downhill naturally unless there is constant intelligent application as in the creations of alternating current which is naturally nonexistent, but intelligence has created it. There will be more on this in the next chapter.

Before I attempt any physical details I would like to review the underlying intelligence that has guided the philosophy of this thesis. Earlier I introduced terms of reality that represented certain laws according to a particular number. In the subjective they were:

    1 defines the LAW of EQUALITY
    2 defines the LAW of DISTINCTION
    3 defines the LAW of PROPORTION

Science used the above in order to understand the elements. In the predicate I introduced:

    4 defines the LAW of CONSERVATION
    5 defines the LAW of MEANING
    6 defines the LAW of MEASUREMENT

These three represent the world of mathematics as far as science is concerned. They show the dynamics or predicate action of existence. The three subjective laws are essential to the three predicate laws and are particular to them. Consider this statement: Conservation is applied across the equation; equality separates the two sides. A proportion is often the distinction of one side and the meaning of a new term is defined by the proportion on the other side of the equation. The whole process is how we measure objective reality provided that individual measurement is equal from one unit to another. The above includes every law and concept without even considering terms for objective reality. It is objective reality that is mostly distorted by science for which it seems to be the masters of. I will explain by starting with the meaning of creation.

    7 defines the LAW of CREATION

Creation is dependent upon all prior terms and numbers. The law pertains only to the principle of organization of the elements whether performed by intelligence or by natural law. The chair we are sitting in is a combination of intelligent engineering of naturally produced by-products. The essence of creation is in the term that generalizes it. In other words creation is infinite thus:

8 defines the LAW of INFINITY

There is nothing more general about creation than infinity. I did not plan the use of eight even though it fits. I originally resisted by trying other arrangements but finely settled upon the obvious. The reason was because of the tendency to read from left to right. I knew that equality was the harmonic and did not start at the left. This I accepted but was insistent on reverting to left and then right as in *4* conservation, *5* meaning and *6* as measurement. I was then reminded that what is in the mind as left or masculine reveals itself in the objective right hand. Creation had to be right-handed. We thus now read from right to left giving the arrangement as follows.

Masculine	Harmonic	Feminine
2. Distinction	1. Equality	3. Proportion
4. Conservation	5. Meaning	6. Measurement
9.	8. Infinity	7. Creation

We could place creation on the left because of the natural crossover from the subjective to the objective, but I would prefer to keep it in the feminine. What then belongs to number nine? Before I suggest a term I would like to discuss what physics has done to infinity. Classically we were taught that geometrically the shortest distance between two points in space was a straight line. Modern physics has placed an intrusion upon our mind by saying the shortest distance is really a curve. Modern physics does this by introducing time into the geometrical theorem. We all know that when we travel upon the earth we travel in a curve. We do so out of practicality and the nature of time but not because a curve is geometrically the shortest distance. A straight line diameter of the earth defines the shortest distance between a point on the equator through the center of the earth and to an opposite point on the equator. Geometry has this ability to conceptualize and define geometrical meanings. Trying to inject time and practicality of travel on the surface of the earth into the meaning of dimension will only annul the meaning of any geometrical term such as diameter. Geometry is subjectively spacial, time is predicated upon motion, and mass is our understanding of objective reality. In terms of geometry the shortest distance is a straight line. In terms of velocity it is another matter. You can fly a plane faster on a larger curve than drive a car on a shorter one. The reason is because we have also introduced the concept of energy as it is related to moving mass against friction.

Relativity attempts to say that space is curved and that if we were to be able to see in perfect detail in any direction we would eventually see the back of our head if we could remain still long enough. This comes from the idea of measuring distance using light and since space is curved, light therefore is curved and folds in upon itself—something like a figure eight. We can continually follow the line of the eight and come back to the beginning. The concept of limiting the size of the universe dates back to mid-evil times when an artist might have described the universe as a very large sphere with stars painted on the inside. The basic problem will all of these scenarios, whether antiquated or modern, is that intelligence does not accept any suggestion that would limit the ability to ask the eternal question, "What is on the other side or what is at right angles to the path of light?" The ability to ask this question requires a positive answer, "More space." This would suggest that intelligence does not curve space—rather it fathoms the infinite. By intimidating the average intelligence that space curves in upon itself is done so in order for the believer to justify their ignorance and convince others they are more intelligence because they can conceive the inconceivable. The modern relativist is simply in denial of infinity. He accepts oblivion as a substitute for infinity in order to prove to himself that he is not afraid of it. Deep down he is terrified. Such was Freud's problem when talking about ghosts.[11]

The ability to conceive of the infinite is intelligence and if you can think of the finite you simply ask what is beyond in order to again comprehend the infinite. Intelligence is greater than infinity. If not, why does intelligence comprehend infinity more easily than measurement introduces a paradox? Infinity is only a particular aspect of intelligence. We thus have our complete terms of objective reality.

9 defines the LAW of INTELLIGENCE.

Perhaps it is premature to talk of the Law of Intelligence. Some would question the Law of Infinity as being the logic of what is beyond any end. We probably need some essence or generalization to work with, in order that intelligence becomes a law. For now it is apparent that science does not consider intelligence separate from biological creation. It seems that science is defiantly set on a track of destroying both infinity and intelligence by making both to be a particular part of objective evolution. Science also considers mathematics objective as well as subjective reason. All things are electric charge and even the magnetic field is only an emanation from electrically charged substance.

Infinity is the harmonic of intelligence and creation. Creation is the feminine aspect as is also the earth. Intelligence is the masculine aspect as is also that of God. There is something unique in the epistemology laid out thus far. We have an epistemology tree that incorporates three realities and within each reality one will find the masculine, feminine and harmonic aspects.

What we have is a Pythagorean tree not unlike the *Kabbalah* or Tree of Life. The harmonics are not only a blend of the masculine and feminine, but the numbers or musical keys 1, 5, 8 and 10 produce a perfect harmonic of a musical scale in any key. This was a perfect coincidence on my part with no intention of achieving this end. The intent was to produce a quantitative tree in order to nail the meaning of reality.

Perfection is what Pythagoras assigned to 10. Perhaps he obtained it from the Egyptian priests holding a common

Chaldean ancestor with the Jews. The Jews had their own terms such as:

2. Wisdom	1. Knowledge	3. Understanding
4. Justice	5. Beauty	5. Mercy
6. Victory	7. Foundation	8. Severity
	10. Kingdom	

One can see, however, that *justice* satisfies the law of conservation, and *mercy* is more like a proportional measurement of man. These are the predicate aspects of reality. *Wisdom* is like distinctive reasoning power and *understanding* is like seeing things in proportion. These are the subjective aspects of reality. There is some correlation, but the numbering varies from authority to authority. One thing for sure is that the Kabalistic tree of life separates reality into three levels—often referred to as the *mind*, *heart* and *loins* or *body*. In the objective level, Jewish creation is the focal point of the word loins. It cannot be assumed that tradition is perfect or that translation from generation to generation produced a perfect tree that mimicked some original after the Chaldean language. Change is understandable—especially from language to language.

The Tree of Life was perhaps an epistemological method of establishing meaning for intelligence to grasp. It perhaps has lost some aspects of this process. Tradition by nature changes meaning and thus the Jewish way may have infiltrated personal meanings of various views not unlike the relativist turning meaning upside down. Intelligent perfection moves "at one" with the universe rather than against it. Intelligence supports harmony rather than attempting to control it. The relativist fails in this whole process by defining objective existence without a verb to describe a perfect action. This is not using intelligence wisely. It should not be assumed that those fostering relativity are unintelligent. It is more a problem with psychology and thus meaning may have had a foundation in religious antiquity in order to prevent the twisting of terms. The relativist changes the usage, measures without tolerance, and breaks the law of intelligence by cheating on conservation. The relativist fails to accept an invitation to generalize and work to the specific.

Significance becomes more important than right action. The relativist speculates in changing existence instead of building a meaningful structure. Observation becomes more important than sound judgment. The relativist and religious counterparts define the action of existence as having a chameleon—like entity that causes appearances and disappearances. They deny the matter of intelligence by assuming nothing beyond it. They insult the intelligence of God, the spirit of faith and the true objectivity of creation. It must be said that science can have the same problems with intelligence that any religion does because equations are used as the scientist's chameleon without explaining according to a level of all intelligence. They are thus able to change the terms to fit a personal reality rather than to the reality intelligence requires. Science needs a Tree of Life and philosophy has failed to provide one.

In essence, intelligence does not create law; it obeys it. The entire world is not a game of arbitrary rules with the people merely players. Universal laws of existence must assuredly coexist eternally with intelligence as separate from the ability to understand from that which is understood. The act of obedience in terms of choice is meaningless without this separation. Existence is real and the intelligent universe must obey. Intelligence has no choice to obey if it is of the same substance as electric matter. Choice indicates error and thus we need a Tree of Meaning. Bound by the laws of the universe, God too must obey or God exists as magic in the mind and less of an existent than that of man who has all three realities we call body, heart and mind. The philosophical corollaries are intelligence, spirit of right action, and physical existence. Does physics have three outcomes? Are they not space, time, and mass? If all this is true then God too must have three basic attributes of intelligence, the spirit of right action, and physical existence.

The relativist denies the objectivity of God and of supreme intelligence over paradox along with action free from paradox. Because of their psychology the relativist stimulates redefinitions and avoids essential generalizations that give objectivity to the magnetic and electric field. The relativist relies

The Einstein Illusion

wholly on observation without reason. To the relativist, seeing becomes existence. Sometimes the physicist will talk of ghost particles such as the neutrino. Although never seen, conservation laws demonstrate that the neutrino is a real particle. When the observation of neutron decay would always yield a proton and an electron splitting in less than opposite directions, the law of conservation of angular momentum appeared to be violated. This is true intelligence applied. It seemed that the existence of a third ghost particle reacting opposite the proton and electron vector was required to save the conservation of energy and angular momentum. The neutrino had many differences from both particles and light, but the relativist's net choice consisted of treating the neutrino like any other particle rather than understanding what it might represent—a possible particle of intelligence or something from another universe. Just like electric matter can be seen and magnetic matter cannot the neutrino is so different that it needs consideration as a third universe. It is this or something totally different than a photon or any photo-electric particle. It is too small to be a conglomerate of matter and too illusive to be a photon. The neutrino was not observed into existence. It did not matter whether it could be detected or not, intelligence has demanded it.

One cannot assume that the photo-electric and the magnetic particle exist as the only basic units in the universe. Closer tolerances and sharper definitions may require a third monopole. In keeping with the theme of reality one can suppose into existence a control-like third monopole. Just how this third monopole exists and how it plays a part may simply be a mystery, but I still wish to attempt a view in hopes of finding meaning for intelligence. If the neutrino is nothing but electric matter, the argument of a particle of intelligence outside of electric matter is still appropriate. Adding a third monopole yields a proportional reality. I prefer to call it a universe of intelligence. The conclusion for this is none the stranger than warped space and dime dilation. It would presuppose that a lot goes on in the surrounding space that the scientist measures as warped emptiness.

When something strange happens, the relativist chooses to explain strangeness in terms of curve-fitting observation. Instead of accepting a higher order between the seen and unseen, the relativist creates immaterial objectivity. Intelligence does not accept immaterial substance. It may be unobservable, but not immaterial. "The wind bloweth where it listeth, and thou hearest the sound thereof, but canst not tell whence it cometh, and whither it goeth:"[12] So it is with every particle of intelligence.

A three-part proportional reality requires a better epistemology that will mimic most poignantly the classic cliché, "There's more out there than meets the eye." God can be part of a complete reality if the generalizations are clear. Just as action at a distance no longer exhibits magic in most forces, we still have the so-called week force that binds neutrons. As we think of protons binding atoms, neutrinos bind neutrons. This is referred to as the week force, something far different than the electromagnetic force described thus far. In this way the neutrino differs from light. Do we have a new form of matter that can pass through the sun quickly without a hit where normal light has to take a relay rout to make it to the surface? Whether the neutrino is a particle of intelligence, or a particle that works with magnetism in a different universe than we know of electric matter, the neutrino is very different, but its close likeness to light is revealing. Our universe is probably right handed—the thumb representing the magnetic field and the fingers the electric field. If this is true even though no one can really determine which way the magnetic field moves, it is understandable that another universe could be left handed to the magnetic field and we may not know of its existence.

In reflecting upon the illusive neutrino that passes through anything and yet it seems to be a control element of the neutron, I wondered like many if it was some sort of photon. The smallest photons can be detected. The neutrino is illusive and carries no charge as does the photon. It is neutral. Many particles are discovered, not because they are seen, but because they are not seen. Their existence is determined by certain laws of momentum and energy. In order to save the laws of conservation things are often suggested and concluded in order to make the laws symmetrical. As suggested in the beginning of this thesis, science must have a psychological problem in not being able to apply intelligent conservation laws to other aspects of reality. The relativity of space, time and mass is a perfect example. God would be another ultimate example.

Science once talked of ghost particles—something real but not seen, but over time the term has fallen into disuse in favor of predictable objectivity. Sigmund Freud had a problem whenever the subject came up about ghosts. Perhaps this was why he considered that mental illness was more like a bad recording in memory. It is not unlike science to look for some strange physical cause rather than admit the possibility of something unseen. There are other terms in reality other than mathematical terms that suggest that measurement can be used in order to develop concepts about equilibrium. In terms of the non physical we often think of harmony and judgment—even good sense and understanding. Does science seems to say. "We must obey a cause and effect criteria and all cause is physical and nothing is unseen." Essentially it is more of an excuse to believe in a paradox rather than find a responsible solution instigated by that which we may not be able to see. It is one thing to say that science operates only upon what can be reproduced in the laboratory, but how many go to great lengths to disprove that which we cannot see. We come to realize that their problem is a psychological one rather than scientific.

A Russian study suggests that the decay rate of radioactive isotopes is not constant as is commonly believed[13] and a recent study[14] also finds that the carbon 14 dating appears to be affected by the rate of neutrinos emitted by the Sun. If true, this finding casts doubt on the absolute reliability of radiometric dating if the neutrino flux from the Sun has not been constant throughout history. I had already considered that the earth's magnetic density may not have always been the same. If it was ever less dense on the surface of the earth, for any reason 100,000 years ago, it would have had a profound effect on atomic clocks as illustrated earlier as well as radioactive radiation. I read the works of an anthropologist that studied the antiquity of the Sumerians. He noted that the carbon 14 dating was not accurate beyond about 4,000 years. I mention this because science often concludes incorrectly—not because of false information, but because of a desire to kill God and escape personal responsibility. It is really a psychological problem in how intelligence deals with the denial of oblivion—a problem rampant in relativity.

Measurements have shown that the number of neutrino types is three. This corresponds to six quarks and six leptons, among them the three neutrinos. However, actual proof that there are only three kinds of neutrinos remains an elusive goal of particle physics. Because antineutrinos and neutrinos are neutral particles, it is possible that they are actually the same particle. Particles which have this property are known as Majorana particles. Several experiments have been proposed to search for this process and determine if in fact they are the same particle, a hypothesis first proposed by the Italian physicist Ettore Majorana. The neutrino could transform into an antineutrino by flipping the orientation of its spin state.[15] This is not any different than what I have indicated with the electron and antielectron. Experimental results show that practically all measurements produced neutrinos having left-handed helicities. It is entirely probable that right-handed neutrinos simply do not exist accept for very short times

The existence of nonzero neutrino mass somewhat complicates the situation. It is believed that neutrinos, like photons, are massless because of their constant motion. A particle with mass is not constant as to linear motion; neither is light according to this thesis unless you are referring to the constant angular helicity as in each photon. I have spent this entire thesis demonstrating the fallacy of this argument. A neutrino just like light can have mass even if it moves near the velocity of light. The calculation for giving light zero mass comes from giving the stationery origin as zero momentum and not angular momentum that is really demonstrated in the helicity of both light and neutrinos. Physicists consider that neutrinos are nearly always ultra-relativistic. This is how they assign mass. If they would understand that the mass of a proton and an electron are less than a neutron, they would understand that the neutrino has mass to make up the difference in the same way light has mass to make up the difference in a smaller electron after emission.

Solar neutrinos originate from the nuclear fusion powering the Sun and other stars, but direct optical observation of the solar core is impossible due to the diffusion of electromagnetic radiation by the huge amount of matter surrounding the core. Neutrinos generated in stellar fusion reactions interact very weakly with matter, and pass through the Sun with few interactions. While photons emitted by the solar core may require some 40,000 years by diffusing into the outer layers of the Sun, neutrinos are virtually unimpeded and cross this distance at nearly the speed of light.[16][17] Neutrinos can travel vast distances with very little attenuation. The galactic core of the Milky Way is completely obscured by dense gas and numerous bright objects. According to some, Neutrinos produced in the galactic core will be measurable by Earth-based neutrino telescopes in the next decade. Neutrinos will be a very useful probe for many important galactic events. Many other uses of the neutrino have been imagined due to its strangeness.

In particle physics the main virtue of studying neutrinos is that they are typically the lowest mass, and hence lowest energy examples of particle physics. For example, one would expect that if there is a fourth class beyond the electron, the neutrino would be the easiest to generate in a particle accelerator. It is believed that neutrinos could also be used for studying quantum gravity effects. Because they are not affected by either the strong interaction or electromagnetism it might be possible to isolate and measure gravitational effects on neutrinos at a quantum level. Because neutrinos are electrically neutral, they are not affected by the electromagnetic forces which act on electrons. Neutrinos are affected only by a "weak" sub-atomic force of much shorter range than electromagnetism, and are therefore able to pass through great distances in matter without being affected by it. If neutrinos have mass, they also interact gravitationally with other massive particles. You can see that physics may be looking in the wrong direction.

If the neutrino is like a helical photon, has mass, is left handed, passes through immense quantities of matter without absorption, yet is a component of every neutron; it sounds very strange indeed. Keep in mind that the very small photons travel in a medium and do not follow a straight line direction as do neutrinos. Large photons can be seen and even larger ones will destroy matter. I consider photons to be right handed. This is opposite the use of the electron as current being left handed, thus current is right handed if it be the photon. If such is the case then neutrinos will not interact electrically as do photons because neutrinos are left handed.

If we take the neutrino as an essential control particle for the neutron and yet we cannot really see it, but know that it must be there, it must be something similar, but opposing in some way to photon binding. If we treat it as something that spirals a magnetic field at rest, but spirals opposite the photon that can be seen and detected we have a left handed universe that plays a part in every atomic nucleus. Perhaps the neutron, for some strange reason, incorporates the smallest unit of intelligence and holds something of both universes—the very smallest piece of electric creation and the smallest unit of intelligent equilibrium and harmony. Our universe has light going every which way and is easy to detect. Neutrinos may be the very same, but we do not see them absorbed by left handed electrons or something very different than our universe. I suspect that when neutrinos combine it is anti-entropic. They move from a lessor order to a higher one but allow the other universe to control this one through the neutron and perhaps other particles if they are so discovered.

If we want to explain ESP, spirit, as well as intelligence there is something physics has discovered by maintaining the law of conservation. If anyone wants to laugh and chuckle at this they can also laugh and chuckle at this writing in hopes of never facing responsibility—the hallmark of intelligence. Consider every laugh at harmony and conservation as the psychological denial of responsibility. God and all intelligence must obey all laws of the universe. To do so is of the most intelligent for any who ascribe to this law. Magic is only a lack of understanding and a desire for irresponsibility. Only through these immutable laws can intelligence ever change the order of things. God did not create these laws. They would have to exist eternally with God. Looking for a God of magic defies responsibility and the law of conservation, harmony, equilibrium, balance, and intelligence that holds all things together.

Science has done well in determining objectivity especially in terms of measurement. Today modern science has not only been able to determine the different elements but can give the precise size of each atom. The inherent problems outside of measurement are that science cannot explain how things work. There is talk of electrons falling into atoms as the positive force of the added protons pull the electrons ever so close and atoms of larger mass often become smaller in dimension. There is eventually a paradox as to why the electron just does not fall completely.

There are many paradoxes not only in relativity but also in the nature of charge and even elementary things like electricity. I think these paradoxes are byproducts of a failure to study field theory and learn more in the measurement of the magnetic field. The atom for example is never discussed as if the magnetic field density plays a role in particle behavior or even basic principles of chemistry.

Samuel Dael

11. Maxwellian Space

> It is indispensable to first dispel a few errors under which electricians have labored for years, owing to the tremendous momentum imparted to the scientific mind through the work of Hertz which has hampered independent thought and experiment.
>
> Nikola Tesla [1]

What is a Radio Wave?

Nikola Tesla considered that there were certain assumptions about electromagnetic radiation that were not true. Tesla perhaps would have a different view other than what follows, but it does express many of the troubling assumptions about the wave properties of light that are in conflict with particle characteristics and even basic electromagnetic theory. The beginning of the assumption begins with a fact: *A static, unvarying electric field will produce a static magnetic field. Also a varying electric field produces a varying magnetic field. The reverse is also true. A static magnetic field produces a static electric field. If the magnetic field varies, so too will the electric field.*

Essentially the concept above leads to the assumption of electromagnetic wave propagation often referred to as having a complete range of frequencies where each is traditionally assigned a particular quantity called hertz (Hz) named after Heinrich Hertz.

As long as one understands that a static electric field means a constant unchanging intensity or direction and that this unvarying electric field produces an unvarying or constant magnetic field the unvarying in this sense means constant motion without a change in direction. The variation in electricity is the essence of alternating current. Static electricity as used above is a direct current without a change in direction. Variation then is a change in direction. Because of a change in direction we can produce a mathematical wave but that tells us nothing about the physical makeup of an electromagnetic wave as we can depict in water or air.

The term electromagnetic wave is very misleading, especially when we see a line curve on a monitor. We fail to realize that it is a mathematical representation that happens to look like a water wave. Objects do not move in a wave form as if they are taking a roller-coaster ride on a wavy track. Objects move up and down or back and forth. They have a constant change in direction and are not static as if moving in a straight line like a photon. The change in direction is what produces the wave and not some constant motion in one direction. The wave can be seen, and has a constant linear motion, but the particles causing the wave are changing direction in localized segments. For the most part this oscillation dissipates when the energy dissipates with a change in medium. This can only happen if there is an abundant of objects bumping into each other that are actually forming a medium. It is the contact that passes the energy. The line curve is only a mathematical depiction of the movement of energy from object to object. We see a water wave because of gravity, the earth's rotation and other forces, but the water molecules do not wave horizontally until they hit shore where the energy is dissipated. This is an illusion. Every objective molecule can pass energy from one particle to another. It all depends upon the medium created. A mathematical wave whether changing water molecules, air molecules or electric particles is the same in principle. The only difference with electric particles is that they act like a cue stick and each and every oscillation of the billiards represent the magnetic particles moving at right angles. A rocking boat could also be compared to and electric field. The energy then travels in the water as it does in a magnetic field as a wave until another boat (electric field) rocks to the same rhythm. Both the electric and magnetic fields are different mediums that affect each other. An electromagnetic wave is only a mathematical convention that assumes that both the electric and magnetic medium is constantly exchanging energy. There is no way to prove this until you understand what an electric field is composed of.

There are two problems with the unification of visible light and electromagnetic wave or radio propagation. First, the magnetic field is not considered to be a physical medium. And second, the electric field lacks understanding. Ever since Maxwell created his equations as a mathematical resolution, we have failed to give the two fields any form of particle nature, yet we consider them to pass energy without substance to carry the momentum. Without a particle to carry the energy and its ability to establish a medium to measure a wave, why call electromagnetic radiation a *wave*? Something cannot have momentum or energy unless it has substance and that it is able to form a medium. We are back to the same problem that began with Maxwell's ability to create an equation for Faraday's magnetic and electric particles as they formed a medium and thus a wave, yet Maxwell denied their physical existence. It is like saying that an electron looses mass when it fires off a photon, but this photon has not mass. The equation does not prove this. It is only a concept the relativist delights in. Without light being a particle or a wave in a medium, modern physics has introduced an existential crisis and Maxwell started it. I would love to discuss this more, but it would take a book on psychology to describe the motivation for denying physical reality as if things move, but there is no object. This is the essence of Maxwellian space.

I guess it is perhaps due to our strange understanding of energy as if it can move outside of substance. We attribute this mystical entity having many powers which we label as gravity, charge, and radiation. We talk like a meta-physicist assuming that mathematics proves the objectivity of such entities, yet we deny the power thereof. We conclude that something moves with zero mass. How we think that nothing can push something

Maxwellian Space | 244

is beyond reason. It is a very religious concept of magic. The failure of this view is manifest in the obvious fact that electrical engineers cannot explain what current is without resorting to the movement of electrons. They are not taught that photons also have mass and thus carry the energy in the opposite direction of the unstable electron. The common understanding of energy is more like studying a rifle's kick and neglecting the kinetic energy of the bullet until it hits something. We measure the beginning and the end of events and assume something strange in-between. We envision the rifle kick of each electron loosing mass as it moves closer to the atom or changes its bonding characteristics and neglect the higher speed photon of a particular voltage moving from electron to electron within a wire until it lifts an electron from its bond at the other end. We mistakenly think electrons moved the entire distance from a negative potential to a positive rather than two separate electrons moving at each end as one picked up a photon from the electrical current initiated by an overly charge electron at the other end.

This psychological ineptitude of not being able to understand current as the movement of photons is a problem with electronics today. The same applies to Modern Physics. When physics talks of energy in light it is like using only the verb action void of an object that moves. We have a subject that sends out light and an object that receives but there is no substance passing between. We call it energy, but we only know that when the light hits something. We assume it had no mass between the fire arm and the target. We know that light is not energy in a medium so reason says it has to be mass. Modern physicists dwell in magic—a mysticism of their own religion called massless electromagnetic light rather than a photon particle of mass.

Mathematics has become objective process in physics, but what is the math really saying? It says that energy is mass times the velocity squared. Zero mass means zero energy. If light then has zero mass, then it also has zero energy. Something cannot have energy if it has no mass. Either the energy is exchanged through contact between mass particles (a medium) or it travels the entire distance with mass. Light therefore has mass and kinetic energy together. In desperation to prove the mysticism of energy, the mathematician creates a graphic depiction of an oscillating motion of substance and plots it on a graph by giving time a dimension. This form of geometry agrees with the math, but not with reality. Water particles do not move horizontally with a water wave. The wave is a graphic illusion, for the water molecules move up and down passing energy from one to the other because of the medium. The wave is a pictorial illusion of mass moving horizontally. It does tell us of the smooth energy exchange, but not of the actual movement of mass particles. So it is with a monitor that plots the up and down motion of an electric current. The programmer also gives the horizontal dimension to represent time and thus we think light waves when it is only the alternating current in a conductor moving back and forth as water moves up and down.

    With a massive motion of electrical current reversing directions rapidly in a wire, it also affects magnetic space within and around the conductor. If magnetic space is a medium, then energy can pass from one point to another as radio radiation. Individual magnetic particles do not travel the distance as does light particles are fired from an electron or atom. That is the difference between radio waves and light. Just as a medium of photons affect a medium of magnetic space within the conductor, we have an exchange of energy that radiates from the broadcasting device at right angles to the wire. The spiral motion of each and every photon is probably consistent in some way to the direction of the rotating magnetic field, thus the right angle relationship may not be as strange as formerly considered. It is the energy that radiates perpendicular to the conductor, but neither the photons nor the magnetic particles travel in that direction. It is the energy passed from one magnetic oscillation to another that travels the distance. Time is only a graphic depiction as illustrated horizontally in **Figure 11:1**.

    The electric field oscillates within a conductor and the magnetic field resonates to it at right angles. The linear depiction

Maxwellian Space | 246

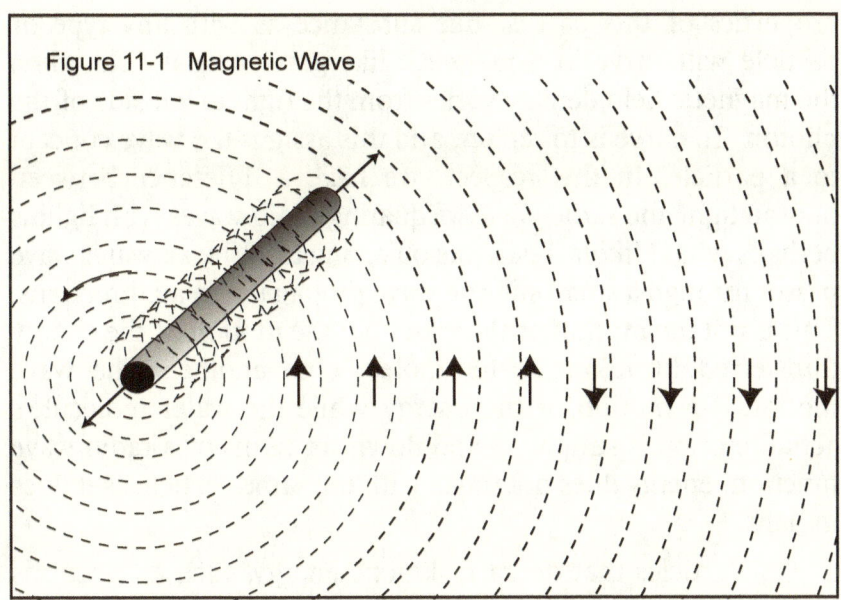

Figure 11-1 Magnetic Wave

outward is not the movement of any particle. This direction depicts only time. The individual arrows of each magnetic layer depict an oscillation at a single location equal to the oscillation in the conductor in past time. Infinity in the equation is a time differential and does not depict substance transcending space. It depicts energy exchange in a medium. The movement of both fields respond infinitely in time but no photon moves the entire distance. The energy, however, transcends space as it radiates from the conductor passing energy from the oscillation of one magnetic area of space to another until a conductor is placed at some distance resonates to the magnetic wave. The electric field exists in two places only at the broadcasting conductor and the antenna conductor. Both must be grounded for the current to flow. Any electric field in-between and not grounded will not resonate. Even if you had a piece of wire in free space no electric current would appear. There is no electric field between any two conductors.

When we talk of an electron spiraling off a photon we do not think of a single curve other than the photo-electric field. We think of a particle-like photon moving in a straight line to be absorbed by another distant electron in need. The wave properties of this particle-like substance as with any type of particle will curve in a magnetic-like gravitational field when the magnetic field density varies from the right to left side of the photon. To curve is to refract, and this assigns the wave concept to a particle. In this respect, we have a difference between photon light and radio waves. Equating radio waves with light is perhaps what Nikola Tesla was objecting to. When a water wave passes through a small slit, the wave propagates in all directions. This is not the motion of the water but the motion of the energy transferred from one water molecule to another. The wave motion is a motion of the energy while the water molecule's actual motion is simply up and down. In terms of a radio wave function, energy does not travel with the same particle as it does in light.

Particles that do carry kinetic energy, such as electrons and photons, turn because there is a sudden change in the magnetic field density in the surrounding space. This is the case when particles pass through a slit. The surrounding magnetic space next to the side of the slit is what causes the curvature. The edge of the slit has more dense magnetic properties and thus causes the photon or any particle to turn more, thus the energy moves with the photon rather than an exchange from magnetic particle to magnetic particle as in the case of radio waves. Just because energy in water and photons refract in a slit does not equate them as wave propagation in a medium. In one case the water particle does not traverse the total distance while in the case of the photon it does. Physics needs to equate radio waves with water or any medium and not with light. All things have a refractive index, but that does not equate them.

The whole reason for this chapter is to resolve the wave particle duality. No one to my knowledge has done this. What is assumed as a wave is simply a change in direction of a particle due to the magnetic field properties adjacent to matter. Despite this fact, physics still talks of wave propagation as in radio waves having a longer wave length than say photons as *x rays* or *gamma rays* produced by the breakup of electrons. This wave length is a length in time and not position as to particles and a length in position as to energy in a medium such as water, air, and the magnetic field. There is a difference and the unification of all wavelengths into the electromagnetic spectrum is just not true—something I think Tesla may have been trying to say.

When Einstein adopted Maxwell's equations he of necessity adopted Maxwellian space. This new form of space was neither the motion of particles nor was it the motion of waves in a medium. It was the propagation of the electromagnetic relationship into infinity. Now that sounds very interesting, but also very misleading.

Maxwell's equations are a result of Faraday's electric current through a wire and the measured results of the surrounding magnetic field. One could perhaps measure the distance between the wire and a magnetic compass and determine the time it takes to energize the current and for the magnetic compass to react. This is simplistic, but considering the time in which a magnetic field reacts to the change in direction to an electric current is essentially the issue. The further away the device of measuring the change in magnetic direction the longer it will take. Einstein based his equations on the time differential and came to the velocity of light as equal to the electric and magnetic exchange over a given distance.

We can also look at radio waves from a different approach. If a current changes direction sixty times in a given second and the surrounding magnetic field changes direction also sixty times in the same second and that the reaction time between the center of the wire and a given distance is equal to the speed of light we can assume that over a distance of 186,000 miles (light = 186,000 miles per sec) we could break up that distance into 60 special segments of 3,100 miles each and notice that the magnetic field would be at rest ready to reverse every 1,550 miles. Few comprehend that the same would happen with the current in the wire that initiates the broadcasting signal. In addition to this it would take an antenna 1,550 miles wide to receive such a signal. I illustrate this in order to demonstrate what is actually going on with a 60 Hz frequency. We now have to ask, "Where is the electric field outside the wire to receive a push at right angles to the magnetic field? Is it every 1,550 miles or even 3,100 miles? If the reaction is just outside the wire as well as 1,000 miles away in less than a fraction of a second, what causes the energy to move from one magnetic level to one 1,000 miles away? In classical terms it is a ripple in magnetic space (aether) where one magnetic shell causes the next to move with an equal energy outward. There is no electric field until you reach the antenna grounded to a receiver. It is this physical conductor and not any so called electric field between that will receive the signal. The equation using the center of the wire to any distance will yield the same results given the velocity of light.

It makes more sense to conclude that radio waves are ripples in magnetic space and photons are small quantum particles fired from electrons or from the center of atoms. The duality of radio wave and electromagnetic particles raises serious questions just as the duality of wave and particle light. No one seems to attempt to resolve these conflicts. They simply accept Maxwellian and Einstein space and the many paradoxes it fosters.

In order to understand Maxwellian space we need to now take a step backwards and not jump to the above conclusion so quickly. Keep in mind that Maxwell's equations describe the electromagnetic relation at any distance between the two fields measured—the one being an electric current through a wire (stream of photons) and the other being the magnetic field surrounding the wire. This means that you can put an electric field anywhere you want just as you can put an antenna at any distance as long as the intensity is strong enough. Does this mean that an electric field exists outside a conducting wire? If you say yes, then you have no concept of what an electric field is other than some force that you measure when you place a device with a current in a wire to measure that force at that particular location. The device itself becomes the source of the electric field so why assume there are an infinite number of electric fields in space between the receiving conductor and the broadcasting conductor? That is essentially what Maxwellian space is. It assumes that both the electric field and the magnetic field permeate space and are not physical in nature. In fact Maxwell considered them both manifestations of the same thing.

The Maxwellian assumption comes from the idea that electric fields emanate in space from charged particles or can be generated out of nowhere by the movement of magnetic space. There is no way to prove this unless you place a conductor in free space. Since the conductor has electrons that can exchange photons, creating a current does not prove that there are free photons at rest in space ready to be accelerated by the presence of a magnetic field. This would contradict the nature of light. If you think the electric field differs from the movement of photons, you are going to have to explain what current is.

The fundamental problem is explaining the nature of vertical forces that radiate as one depicts charge and gravity. We know what a magnetic field is but Maxwellian space tries to create an electric field having vertical characteristics like gravity as if energy emanates and pulls things to it. **Figure 11:2** illustrates the different concepts.

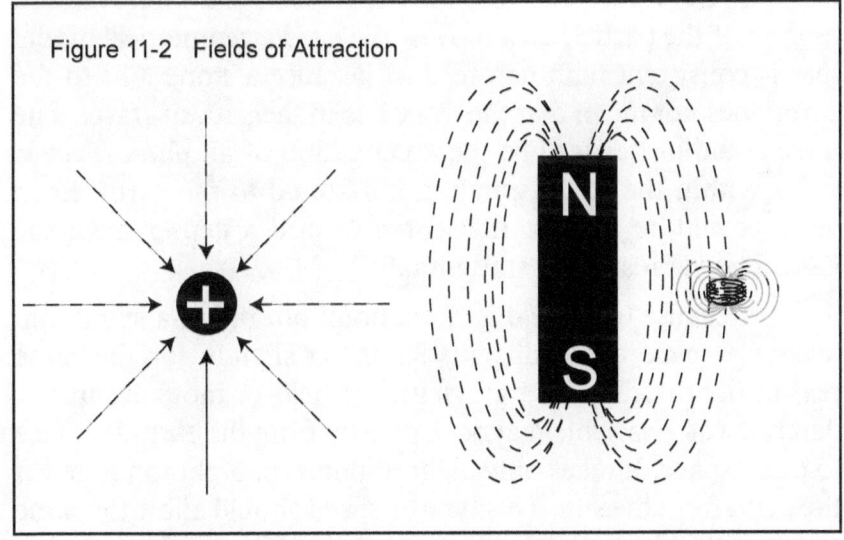

Figure 11-2 Fields of Attraction

The positively charged particle is nothing but a polarized miniature magnet similar to an iron filing having a magnetic field yielding north and south polarity. The positive could represent the North Pole looking down. The arrows of attraction are nothing but the direction of movement, but the cause is the same cause as gravity. The difference is only in the graduated density difference. If the magnetic density becomes closer within a given distance the force would be greater. The smaller particle aligns with the magnetic field and is attracted to the magnet. In the case of iron filings gravity is stronger at certain distances, but eventually the magnet becomes stronger as the polarize particle gets ever so close. If the particle is a molecule with poles running in every direction, the magnet will have no effect unless the molecule is floating. We might be able to study magnetic density by studying gas molecules and how they behave.

If we conclude that some form of energy emanates from a particle and pulls things in, you have to demonstrate or explain it in geometrical terms. You have to correlate it with other aspects of nature and not simply conclude by observation and the math that is set to agree with it. Observation does not explain why. The mathematician may conclude that perpendicular lines are patterns that objects move along, but that does not explain the cause of the force. The line only represents the direction and motion of the particle as it moves. We need a geometrical model that is consistent with nature. Just because a stone falls to the earth does not mean that the force lies in the path of travel. The force could just as well be the acceleration of all photo-electric fields within the stone spiraling horizontal to the earth. Each vector could be a horizontal curve toward a denser magnetic field. The net result is a straight fall.

Because each field is not a point but rather a spiral that takes up space, it would curve ever so slightly for the same reason light refracts in a gravitational field or more accurately described as a variable magnetic density from the earth's surface to outer space. It takes only eight minutes for a photon to reach the earth from the sun. This type of speed should allow the same structure of photo-electric properties in matter to place tension on magnetic properties in space. If the physical magnetic density diminishes from the earth then the tension should be a slight horizontal curve. Since all electric fields are pulling in all directions there is a net curvature to earth due to the tension of a greater magnetic density. Once the object begins to fall, its inertia is added to a little more tension at the rate of an infinite number of cycles per seconded. Action at a distance does not come out of the earth. There has to be a physical, local, material cause. The force of this cause is not perpendicular to the earth. The force lies horizontal to the earth in the form of individual electric fields chewing up the magnetic field and nudging to a denser region upon each cycle at the speed of light.

Maxwellian space adds a mystical conclusion not in a classical view of a magnetic ripple. Maxwellian space denies the existence of physical magnetic properties that can carry a ripple of energy. In addition to empty space and the appearance of an electric field out of nowhere, this imaginary exchange is assumed throughout the immensity of space as both fields appear from nothing constantly exchanging energy. A classical view of radio waves would say that the electric field only resides in the transmitter and the receiving antenna and not in-between. The physical magnetic medium produces the ripple in space.

If we insist upon an eclectic field without a current at many points in-between we conclude without proof, for as soon as we insert a wire providing a closed circuit we catch the ripple and in no way prove the existence of any electric field in free space other than as a component of light traveling unaffected by a radio wave. Just because a conducting wire is placed at a receiving point in the form of an antenna does not assume that there were an infinite number of electric fields between the transmitter and receiver in order to exchange the energy. There is no reason not to suppose that the wave is an oscillating wave in the magnetic field only. If the magnetic field has physical properties such as has been illustrated, classical dynamics would easily suggest such an event.

James Maxwell (1831–1879) followed Hertz with a classical electromagnetic theory, synthesizing all previous

The Einstein Illusion

unrelated observations, experiments and equations of electricity, magnetism and even optics into a consistent theory.[2] His set of equations predicts that electricity, magnetism and even light are all manifestations of the same phenomenon: the electromagnetic field. Maxwell had studied and commented on the field of electricity and magnetism after Faraday introduced his *"lines of force"* to the Cambridge Philosophical Society. Maxwell had calculated that the speed of propagation of an electromagnetic field is approximately that of the speed of light. He considered this to be more than just a coincidence, and commented *"We can scarcely avoid the conclusion that light consists in the transverse undulations of the same medium which is the cause of electric and magnetic phenomena."*[3] This is the wave phenomena that seared the study of light. The velocity and mathematics are correct, but the geometry that followed is not.

Working on the problem further, Maxwell showed that the equations predict the existence of waves of oscillating electric and magnetic fields that travel through empty space at a speed that could be predicted from simple electrical experiments; using the data available at the time, Maxwell obtained a velocity of 310,740,000 meters per second. In his 1864 paper "A Dynamical Theory of the Electromagnetic Field," Maxwell wrote, "The agreement of the results seems to show that light and magnetism are affections of the same substance, and that light is an electromagnetic disturbance propagated through the field according to electromagnetic laws."[4]

Maxwell's work in electromagnetism has been called the *"second great unification in physics"*[5] after the first one carried out by Isaac Newton, but Maxwell had no geometry to illustrate the workings of his mathematics other than lines of force. The lines were mathematical depictions of the movement of energy, but not the force. The geometry came from Faraday's examples of real magnetic particles. Faraday never suggested empty space and even Maxwell assumed *"affections of the same substance."* At the time Maxwell developed his equations, he believed that the propagation of light required a medium for the waves, dubbed the luminiferous aether. The existence of such a medium that permeates all space and yet remains apparently undetectable by mechanical means, proved difficult to reconcile with experiments such as the Michelson-Morley experiment. The basic problem was that the aether was considered ridged rather than variable as the magnetic field. If they equated the aether with the magnetic field, we would not have had the problems and the paradoxes we have today.

If Maxwell and Einstein were wrong and the aether wind was nothing but the magnetic field in varying densities, both will have to change their geometry of space and their epistemology of time despite the fact that the equations work in agreement to observation. Einstein created empty space even after Maxwell's initial reference to "substance." Every physicist since has assumed as much and continually has had trouble making some geometrical sense of the electromagnetic relation. Like Maxwell they think of immaterial imaginary forces of electric fields. Maxwell's equation is the relation of the alternating current in a wire to the magnetic space outside the wire. It does not go beyond that. The reaction time between the two fields, given a specific distance, calculates to be the velocity of light between the two fields. The mathematics may carry it to a continuous wave function, but the equation is based upon the isolated relation of the electric and magnetic reactions at a specified distance. Infinity is only an extension of a local event by giving time a dimension.

Radio Antennas

Radio signals are at a frequency that cannot be heard by the human ear and they are so weak they cannot be detected without an antenna and radio receiver. Also, radio signals travel in the form of waves thought to be the same thing as light, but obvious arguments suggest a difference. As sound travels in waves through molecules of air, water, or almost any material, radio waves can just as easily travel in a magnetic medium using electric conductors. A conductor is an electromagnetic device that collects or emits radio waves. It consists of a material that conducts electricity alternated in such a way that it creates a radio frequency in the magnetic field and in tune with the alternating current within the conductor. Like one tuning fork in the presence of another, one can pick up the same frequency broadcast by tapping the first. An antenna tuned to a particular frequency will resonate to a radio signal of the same frequency broadcast. When properly tuned, an antenna will collect this energy and make it available to drive the amplifiers in a radio receiver making it possible for the human ear to hear the same sound expelled into a microphone and broadcast into magnetic space. Two electric conductors are illustrated in **Figure 11-3**.

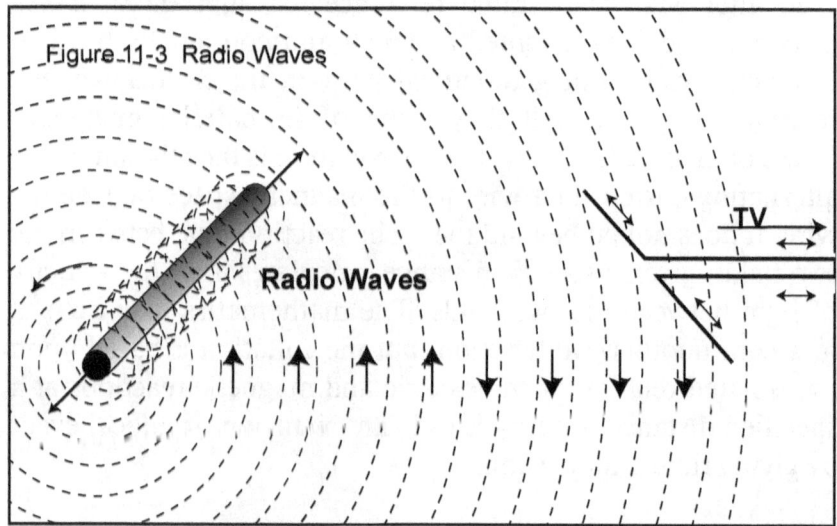

The drawing is simplistic knowing that it is difficult to draw a three dimensional concept. Also, broadcasting conductors come in different shapes creating a specific intensity and direction of a desired signal. It might be hard for some to put physical properties to a magnetic field, but you have to in order to conserve momentum. If you prefer, you can say metaphysical, but you have to understand that the two fields impinge or react to each other. That is the essence of their reality. What makes the magnetic field so unique is that regardless of its churning and dynamic oscillation, it does not affect the velocity of light unless there is a change in density in the path of travel or from side to side causing light to curve. In the case of radio waves in a magnetic medium, electric density would affect the velocity of a wave. This would not be so much because of the electric particles, but because of the magnetic density itself imbedded within electric matter. Magnetic waves would only be affected by density in the same way that light is. This is why particle light (electric spirals) and radio waves (oscillating ripples in the magnetic field) are easily confused—they both refract and travel at a velocity relative to the density of the magnetic field. The difference is that the photon is not dissipated and colors can be separated, but a radio wave will dissipate in time within electric matter but will none the less pass through as light is basically reflected. Radio waves do not reflect. What is generally thought of as reflection is resonance where an electric field in a conductor resonates to the frequency of the magnetic wave. The electric current then causes a new magnetic wave to move in a new direction such as found in a satellite dish. Light can heat up and damage electric components but magnetic radio waves will only affect the direction of the current. Large size photons are more like bullets that penetrates flesh, but might be stopped by a bone. The ability of large photons to destroy is due to their higher mass that physically breaks up electron bonds. Waves pass through a wall because the energy is carried from magnetic particle to magnetic particle. Only if the electric density is great is the wave absorbed or staticized with the massive density of many magnetic fields going every which way.

In basic free space we have an oscillating magnetic field at any point in space changing directions at a specified number of times per second, but the magnetic particles in the first shell or circle never travel the distance like light. Only the energy passes from one adjacent layer to the next agitating one magnetic shell to another. Unlike an electric field that causes a magnetic field to move at right angles, a magnetic field causes its own properties to move in the same direction until the energy is reversed by the electric field in a conductor. Note that the magnetic field changes direction many times before it hits the antenna. This depends upon the frequency of the change in direction of the current in the broadcasting conductor. For maximum reception the antenna has to be the length of two points at rest from crest to crest. It can be one half or less, but the reception is poor if insufficient magnetic lines pass through the antenna in order to energize the grounded electric current. The quality of the signal received is contingent on the antenna, but if the electronics can filter and amplify efficiently, then antenna designs can be hidden within the receiving device.

Once each photon is ejected from an electron it will travel eternally until absorbed. Radio waves do not manifest this characteristic. When a photon leaves an electron it does not radiate in a total circle. It has a specific path for each photon. In this sense, radio waves truly radiate. Photons only radiate in numbers and each has a particular voltage or size that refracts to yield a specific frequency. With radio waves the frequency is the change in directions of each magnetic shell and not a point-like particle. Radio energy is a wave in a medium. A very small infrared photon arising from a warm surface could be a specific voltage we have assigned radio waves, but the two could none the less represent different electromagnetic relationships. It does not take a special antenna to pick up infrared photons. They can be refracted and focused just like visible light, but radio waves work on a different principle. Radio waves can also be reflected. This is essentially what happens in a satellite dish. The frequency does not change, but it is intensified at the LNB because the electric current in the conductor focuses the newly created wave onto the LNB receiver. Radio waves do not pass through metal as easily as through residential walls. In this case they seem to reflect off the dish light photons, but only cause the current to react and stimulate a new magnetic wave in what one might think as a reflective orientation.

Now consider every photon in a conductor. Most would not remember the days of the rabbit ears on a TV. They worked in much the same principle as an antenna does on the roof. The difference is that you had to extend the antenna out according to the particular range of TV channels. If you were to pick up an

HDTV rabbit ear antenna for today's high definition or digital TV, the very same principle is required. Most did not understand that the degree of extension produced the best reception and not so much the angle one would fiddled with. The distance of a certain extension is designed to receive most of the traditional VHF TV channels such as 2 to 7. Shorter lengths are designed to receive shorter wave lengths such as UHF. We are apt to think of a single curve when we think of wavelength, but this is not the actual geometrical representation. A single curve is only a mathematical convention converted to a geometrical layout by giving time a linear x axis and the oscillation is placed on the y axis.

The real frequency is determined by how long it takes electrical current to move from one end of an antenna to another. If the alternating current in a transmitter changes directions at the precise moment a photon reaches the end and again when a photon reaches the other end, we have a frequency of alternating current specific to that antenna or conductor. What we have is the velocity of light and the frequency of change in direction. When you divide the velocity of light by the frequency in change, you come up with a particular geometrical length and not a particular wave. If we could watch a single photon move from one end of a conductor to another, the time it takes for the photon to make a complete path from one end to the other is equal to the time the broadcast magnetic wave changes direction.

It is the change in the direction of the electrical current that produces the magnetic wave and not a static direct current. Simply divide 300,000,000 meters by 92,500,000 (92.5 MHz) and you get a conductor that is 3.24 meters long. In other words, a photon would travel the length of a 3.24-meter conductor before the direction of the current would change. The photon will then move from one end of the conductor to the other 92,500,000 times per second at 92.5 MHz. Photons are used because electrical current mimics the velocity of light, and as explained in this thesis—electrical current is the movement of photons.

The length of a conductor will become strongly excited when photons move in phase and match the time it takes a photon to travel twice the length of the conductor. This would be equal to one cycle. This illustrates the concept of a ½ and ¼ wavelength antenna. In order to be in phase the antenna needs to be ½ or ¼ the wave length or the time it takes a photon to move from one end to the other in phase with the magnetic ripple. Take the 60 MHz. A ½ cycle or wavelength would be about 93 inches and a ¼ cycle would be 45 inches. The size of an antenna must be either the full length, half or one quarter to be in phase. Take a typical conductor on a TV antenna—the one usually attached to the Leeds. This conductor is about 90 or 45 inches because channel 2 is about 60 MHz. This means that a photon can travel the distance in phase within the electrical current moving at right angles to the magnetic field. TV antennas are actually a little more complex but the concept is clear:

> In the US, a TV channel is a 6 MHz wide chunk of bandspace. The bottom edge of the over-the-air channel 2 is 54 MHz and the upper edge is 60 MHz. Within this 6 MHz space is a video carrier, a color carrier, and an audio carrier.
>
> The frequency of the video carrier is 1.25 MHz above the lower edge, so for channel 2 the video is at 55.25 MHz.
>
> The color carrier is approx. 3.58 MHz above the _video_ carrier (N.B. not the lower edge), so for ch 2 it is 58.83 MHz.
>
> The audio carrier is 4.5 MHz above the _video_ carrier, so for ch 2 it is 59.75 MHz.
>
> The reason (for) all of this is that you have to be careful to note how a channel frequency is being specified. The two common methods are (1) specifying the lower edge (e.g. 54 MHz, but often as a range of lower edge to upper edge like 54-60 MHz) and (2) specifying the frequency of the video carrier (e.g. 55.25 MHz).[6]

If a conductor is split at the center where two leads are attached, it does not affect the result because we are not concerned about the current in the leads—only the conductor being in phase with the magnetic wave. Just because a photon moves in the conductor and down the leads, does not affect the magnetic field's ability to be in phase with the length of the conductor.

The dipole antenna is so named because it has two electrical poles, not two physical extensions; it also has two zero points as far as the velocity of the current. This is the point at which current changes direction. When the length is such that the poles are at the ends of the conductor and the zeros are at the center, the antenna will be exactly 1/2 wavelength long. Therefore, a dipole antenna is exactly 1/2 wavelength long. A dipole is most commonly fed at the center, where it presents a pure resistive, balanced, 68 Ohm load to the feed line. A dipole can be fed anywhere along its length, however *center fed* and *end fed* are the most common. If they are fed at the center, the zero rest point or change in direction is at the ends.

The relation of MHz to Hz:

1 Hz = 1 cycle per second
1 MHz = 1 million cycles per second
60 MHz = 60,000,000 Hz

Light = 186,000 miles/sec
1 mile = 63,360 inches or 11,784,960,000.00 in/sec
60 MHz = 60,000,000 cycles per second
Full antenna would be 196 inches and ½ would be 98"

We have to imagine that a mathematical magnetic line is passing through the conductor and is affecting the electrical current. When the magnetic field changes direction so also does the current. Radio waves are far more complex than illustrated. It is almost as if each magnetic field can carry a great many variations in its motion but not direction. At a given moment all magnetic lines are running in the same direction but vary slightly in velocity. This is due to the embellishment placed upon the broadcasting conductor. Technology has developed to such an extent that variations can be digitized. In this way a special receiver can filter or lock a particular variation. There is no such thing as a digital antenna. It is the electronics that separate or digitize the physical variations.

At a given MHz the magnetic field reverses direction and in turn the current in the antenna reverses its direction. All lines would be moving in the same direction over a distance of

The Einstein Illusion

about 196.4 inches at 60 MHz. At a specific distance of 196.4 inches or 1/60th of a second the magnetic field appropriately changes the direction along with the electrical current in the antenna and the broadcasting conductor. In fact these lines lie perpendicularly along the entire length of the antenna and beyond. It is reasonable to assume that it is the magnetic lines passing through the conductor that determine the direction of current. Once the trough or crest passes, the current is reversed. Each magnetic line varies in velocity depending on how close it is to the point at rest. There are also variations in-between depending upon the broadcasting conductor's effect on the magnetic field.

As the current moves back and forth in the conductor it does so in the same frequency as the magnetic field moves up or down. There are many frequencies available and the typical TV antenna can pick up channels 2 through 7 without any difficulty. Shorter conductors placed at an angle are added for higher channels. Other elements are added to reflect, such as the conductor farthest away. Others reinforce the signal. This would mean that many frequencies are conducted through the Leeds, but the tuner only accepts a certain frequency depending on its tuning. This is basic to the principle of TV tuners.

It is most interesting to see that magnetic space can carry almost an infinite number of frequencies in all directions as does light carry an infinite number of photon sizes in all directions. A lot goes on in magnetic space where a single photon is undisturbed and a radio frequency is also practically undisturbed without interruption. The question arises: do we really need an electric field outside of the conductor if a magnetic wave created by one conductor can propagate in space until it eventually induces a current in another conductor? If it is accepted that the magnetic field at rest or in motion can carry an oscillating ripple through its variable density, the magnetic field becomes nothing but a classical aether in principle. It was only put away because science was trying to measure something ridged and not something variable with a graduated density surrounding

Maxwellian Space | 262

matter. Magnetic density is cohesive to any large mass and circulates through it but is relatively at rest in outer space. The natural motion of any magnetic field, such as surrounding the earth, would not affect the velocity of light as much in the East to West direction as it probably would in the North to South direction. Experiment should be devised to demonstrate this or prove this conclusion wrong. Atomic clocks in the East-West direction have already demonstrated this conclusion. What we need is a jet airliner to carry atomic clocks from magnetic pole to pole and compare directions. Keep in mind that the reverse direction would cancel out and the time of day would have an effect if the airline changed distance relative to the sun. Even within matter, in the case of radio waves, the magnetic density would carry a ripple and pass right through while visible light would simply be reflected or absorbed and radiated as heat. This agrees with observation and is the basic reason why radio waves and light differ. It can easily be assumed that photo-electric particles are much larger, and less perhaps in numbers, but equal in momentum given a specific amount of magnetic particle that react to a specific amount of magnetic particles. A great number of magnetic particles creating the flux through and axis will move much slower than the velocity of light.

A photo-electric spiral traverses a magnetic field and varies in velocity according to a particular magnetic density. It should be considered that it was Nikola Tesla who did not consider the aether to be a rigid frame and suggested it to be more like a gas that distributes itself somewhat evenly accept within and around mass thus perhaps causing curvature. Tesla's argument:

> ...It might be inferred that I am alluding to the curvature of space supposed to exist according to the teachings of relativity, but nothing could be further from my mind. I hold that space cannot be curved, for the simple reason that it can have no properties. It might as well be said that God has properties. He has not but only attributes and these are of our own making. *Of properties we can only speak when dealing with matter filling the space.* To say that in the presence of large bodies space becomes curved, is equivalent to stating

The Einstein Illusion

that something can act upon nothing. I, for one, refuse to subscribe to such a view.[7]

Note: According to scientific trends rising up since the dark ages, God did not have properties. This was primarily due to mystical concepts introduced by medieval Christianity. The Jews of antiquity and the early Christians did give God physical properties and eternal attributes that man could actually acquire. Tesla's view of the properties of God was contingent on the religious views of the time. It is noteworthy, however, that space had properties and was not really empty. Space is not warped, but light changes direction due to the properties of magnetic space. Tesla also talked of gravity being more dynamic in nature:

> ...Supposing that the bodies act upon the surrounding space causing curving of the same, it appears to my simple mind that the curved spaces must react on the bodies, and producing the opposite effects, straightening out the curves. Since action and reaction are coexistent, it follows that the supposed curvature of space is entirely impossible - But even if it existed it would not explain the motions of the bodies as observed. Only the existence of a field of force can account for the motions of the bodies as observed, and its assumption dispenses with space curvature. All literature on this subject is futile and destined to oblivion. So are all attempts to explain the workings of the universe without recognizing the existence of the ether and the indispensable function it plays in the phenomena."
> "My second discovery was of a physical truth of the greatest importance. As I have searched the entire scientific records in more than a half dozen languages for a long time without finding the least anticipation, I consider myself the original discoverer of this truth, which can be expressed by the statement: There is no energy in matter other than that received from the environment.[8]
>
> Nikola Tesla

When physics assimilated radio into frequencies of the electromagnetic spectrum, it was perhaps done because of radio's calculated velocity of light and not upon the principle of ripples in a magnetic medium or eather. I know of no conductors that broadcast visible photons unless you say that this is precisely what the electron does. Truly then you have a relative concept of mass turning into pure motion of nothing. Energy loses its classical meaning of mass in motion and adopts the relative concept of changing raw mass into mass-less motion. It's like turning an object into a verb. When we see a photon, do we assume an electrical conductor at its origin? Do we consider x-rays to originate the same way? I think not. Certain frequencies are detected in different ways by different methods. This alone suggests the measurement of different realities and not the same medium or particle. A refractive index of a particle is not the same thing as the frequency of photons reversing direction. With particles, we are talking completely different wave length concepts.

It has been illustrated how a conductor can broadcast radio waves. In most cases this process has become more sophisticated with the ability to condense or magnify the energy into a single point such as found in a satellite dish which can both broadcast and receive. It would make more sense to think in terms of photons reflecting off a dish and focusing on the LNB receiver in order to intensify the signal. If such were the case then the broadcast means would have to generate photons from a similar device that would radiate in all directions and when the dish reflected the photons they would reflect more like a large spot-light reflector into the night sky. It is either this or a reflector becomes a conductor that would be electrically excited by the incoming wave and thus generate a new wave in sync to that which is received, but the direction would be changed if the conductor is what has become the traditional dish. It will be as if the conductor focuses the signal in order to intensify it but not change its frequency. We need to revaluate the concept of a medium—more specifically a magnetic medium. We can call it magnetic space that electric matter acts upon. We can even call it aether if you want to go back in time when the mind of man was far more perceptive in understanding than today's paradoxes.

The Nature of Charge

In Figure 11-2 it was illustrated the traditional difference between the magnetic field and the so-called perpendicular electric field as working similar to principle as gravity. Attraction using the magnetic field can be demonstrated and understood because it circulates. But when we think the electric field emanates there is no understanding. It has been demonstrated that gravity is nothing but the individual electric circulating fields, stationery as to linear motion, acting upon magnetic space. There is no graviton that emanates from the earth. Likewise there is no electric particle field that emanates from a charged particle. Although this is usually a way of depicting charge, this section will attempt to illustrate a better way that has already been touched upon.

Consider that the magnetic field circulates and is weaker or slower in motion the further distant. The magnetic field can be concluded, but the one-way electric field cannot be concluded any more than one can conclude how gravity works. The traditional geometry of both gravity and charge is only a mathematical convention. For charged energy to emanate and never return would make sense in terms of light in that it can also be absorbed because it is simply a piece of mass in leaving the electron or atom and adding itself to something far distant. It has been illustrated that light is nothing but angular momentum being converted to linear momentum (mass included) and then again converted back to angular momentum once absorbed? Charge, like gravity, can also be explained using the electrometric relation.

Traditionally, the magnetic field has a north and South Pole. The flux direction depends upon whether the field is moving external or internal to the rotational cause—the electric field. Now fields of electrons move from negative to positive, but this is only concluded in a small, local sense as an electron is attracted to an atom's polarized nucleus or bonding situation. An orbiting electron field around a nucleus or the orbiting crust of the earth is nothing but the movement of photo-electric particle accumulation in an orbit thus increasing the net magnetic flux in both atom and earth.

To think that electrons move completely through a conductor from a negative pole to a positive is the primary error. They do in part, but the electron is not the current. Electrons jostle back and forth between atoms but they do not move at the speed of light through a wire. In the case of charged electrons we are thinking of overloaded photo-electric fields with magnetic poles. Attraction comes as the electron gives up some of its quantum charge to move closer to the atom or change its bonding relationship. Charge here defined is not in the attraction or repulsion due to the magnetic density, but total voltage potential or how much light is crammed into the electron. Even though an electron or any particles is pulled or repelled due to magnetic density, charge is like pumping up electrons ready to explode. This potential is nothing but an electron loosing charge as it moves closer or increases in charge as it moves away from an atom. Charge has nothing to do with attraction, but it has everything to do with the perfect magnetic density for the size of the particle. Magnetic attraction differs only because the poles of two elements differ. The attraction is caused by the magnetic flux and is not due to this concept of charge.

A free electron has a particular charge—the maximum amount of photo-electric fields spinning like a top. This total in a free electron is the measurement of a unit of charge. When an electron moves closer to an atom it loses its charge in quantum units. In other words it loses a particular frequency of light thought to be waves, but it is really photons moving through space or through a wire from electron to electron to be absorbed by another electron. Electrons are charged because they are holding the maximum amount of photons at a given shell level. It is as if they are pumped up in size causing specific chemical reactions in the positive side in the case of a battery. If you have a chemical condition in need of photons within the negative side of a battery and you connect a circuit to the negative terminal, the negative potential will draw photons from the wire in a chain reaction until the positive side gives up photons until a discharge results. Current is the movement of photons and not electrons.

Certain elements such as copper, iron and even aluminum have an abundance of outer shell electrons to exchange a massive influx of photons and expel the same to a ground. Current, then, is the concentration of an electric field of photons moving between electron and electron. The so-called frequency is so low that the electrons do not move from shell to shell, but are simply used as a conduit for the movement of a specific charge. The electrons will vibrate, but will not change shells as erratically as they do in the chemical reactions deep within the battery. In most respects light is traveling in every direction as a particular volt and does not normally affect the exterior magnetic field. Within a conductor the situation differs—a photon current is directional and thus magnetic space external to light's path within a wire is affected.

Within a magnet it is believed that there are an infinite number of electron-like particles inside with spinning electric fields that are polarized in order to accelerate the magnetic field through the magnet. It is believed that the strength of the magnet is due to the alignments of the iron molecules or other metal particles. This manifests the right angle electromagnetic relationship. Two particles with opposite circulating electric fields can share a common magnetic field and thus be attracted to each other up to a point. The reason an electron does not fall into the nucleus is because it is attracted only to the point of magnetic equilibrium. Charge in other words is the maintenance of magnetic equilibrium whether it is photons within a wire or free electrons seeking more magnetic density. Like charges repel only because the magnetic fields are antagonistic to each other in the same way magnets can perform positively or negatively to another magnet. The perpendicular straight line diagram of an electric field is only a convention describing the movement of a particle needing or repelling more or less magnetic density. The direction of movement is the direction of the so-called lines of force, but the cause of charge has nothing to do with the direction of motion. Magnetic density is the cause. Charge is only a mathematical convention for purposes of calculation. That calculation is done with units defined as the total volts of one free electron—called an electron volt.

The conventional description of charged particles often creates a paradox, but if charge is nothing more than finding a relative magnetic density as suggested in the chapter *Action at a Distance,* and if the electric field is nothing but the spiral of photons around a magnetic field at rest or through a conductor as suggested earlier, then there is absolutely no need of charge. Charge is simply a mathematical convention to explain the behavioral direction of the moving particle in the same way that a gravitational equation is used to explain the direction of gravity. Equations do not show the geometry of an event. Equations only predict the action or direction. It takes intuitive knowledge to see the geometrical relationships. Relativity equations do not predict curved space. They predict curved light. This is the fundamental problem with the minds of modern science. They cannot see the geometry or even attempt to illustrate it. They are caught up into the realm of mathematical formulas and lack the ability to specially construct a model for reality. If the equation works who cares what is going on—just make up something and say that the equation proves it.

Electric charge of a macroscopic object is the sum of the electric charges of its constituent particles. Often, the net electric charge is zero, since the number of electrons in every atom is equal to the number of the protons. The nature of electric charge was proposed by Michael Faraday in his electrolysis experiments, then directly demonstrated by Robert Millikan in his oil-drop experiment. He was quite accurate in the determination of the fundamental unit of charge. Millikan and Fletcher's apparatus also provides a "hands on" demonstration that charge is actually quantified. It demonstrates this simply and elegantly. It should be said though that charge is related to the intrinsic nature of mass and not something that emanates external to it. Just as matter measures a certain quantity of light, it also measures a certain quantity of charge. To think of charge as mysteriously something separate from mass is to give properties to charge that should not be given.

Robert Millikan's design is just a uniform electric field between a pair of parallel plates that lie horizontal with a large potential difference. When the oil drops fall between the plates and remain suspended between the positive and negative plates, it is done so by changing the voltage. The oil drops can be made to rise and fall. A ring of insulating material is used to hold

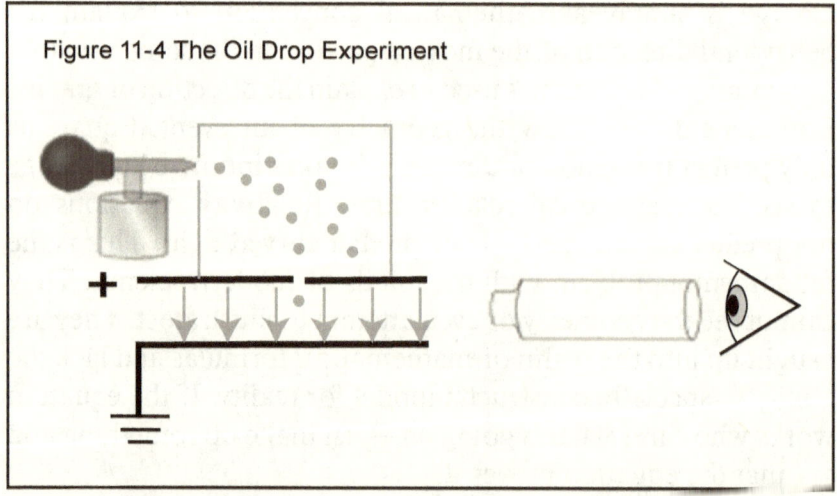

Figure 11-4 The Oil Drop Experiment

the plates in close proximity to each other. The plates have four holes cut into them and three have a bright light shining through them, and the other has a microscope placed at one end to view the droplets. **Figure 11-4** illustrates the apparatus.

The oil apparatus is usually performed in a vacuum because this type of oil has an extremely low vapor pressure. Ordinary oil would evaporate away under the heat of the light source, so the mass of the oil drop would not remain constant over the course of the experiment. Some oil drops will pick up a charge through friction with the nozzle as they are sprayed, but more can be charged by including an ionizing radiation source.

Initially the oil drops are allowed to fall between the plates with the electric current turned off. The current is then turned on and, if it is large enough, some of the charged drops will start to rise. This is because the upwards electric force F_E is greater for them than the downwards gravitational force W, in the same way bits of paper can be picked up by a charged rubber rod. A likely looking drop is selected and kept in the middle of the field of view by alternately switching off the voltage until all the other drops have fallen. The experiment is then continued with this one drop. That is the essentials of the experiment.

It is considered that the electrical force F_E moves from the lower negative to the higher positive and gravity moves in this case downward. Theoretically this would be the movement of electrons upward if current is the movement of electrons. One must ask how an increase in bombarding electrons, moving vertically upward, affects any particle in terms of gravity. By ignoring this we suffer somewhat a contradiction to the classical concept of electrical current that moves from the positive to the negative. Either electrical current is an electric field of photons or it is the motion of electrons or something different. Trying to manufacture an electric field different from one or the other is an assumption. If the electric field is the motion of photons rather than specific to electrons, then the current moves from positive to negative. This was the classical form that did not equate direct current moving from the negative to positive.

The force then may be due to something different than the direction or motion of photons moving from the upper plate to the lower plate. The most interesting aspect is that a higher voltage potential is required to pass sufficient photons between the plates without a conductor. Also, in order to sustain the higher positive voltage or charge, there would need to be a denser magnetic field near the top plate. If we return to the concept of gravity being a graduated magnetic density of increased proportions toward the earth causing electric fields to curve downward we have actually reversed the situation with a greater density above. This difference is calculated in terms of voltage as a force less that of gravity. What we have done is simply neutralized gravity or actually reversed it.

Let us review the process. Photons are going to remain with the electron as long as possible. But when the electron in an outer shell cannot maintain the magnetic flux necessary to hold it together, it fires a very low frequency, or better put, a very small photon in order to maintain equilibrium. Outer shell electrons are the largest of any electron within an atom. Within this valance, they are able to hold more small photons and become excited or in other words charged. The larger size is necessary in order to produce its own magnetic flux and share less with the atom. When an electron or pair of electrons shares their magnetic flux with two atoms, bonding occurs. If you rotate magnets within a wire-wrapped container, the motion of the magnetic field will induce an electrical current of photons in the wire—the more wrappings of wire the greater the voltage potential at the two ends of the wire. With sufficient speed and windings the two ends create a direct current and can light up a bulb of a specific voltage.

You could think of the rotating magnetic field as a means of reducing magnetic density at one point of rotation and increasing magnetic density when a magnetic pole is near an electron. This constant flux causes the electrons to absorb and emit photons and also increase the flow in one direction. If there is no load between the poles, the current does not move but piles up in the positive end creating a denser magnetic surface within the space surrounding the positive pole? The two plates in the oil drop experiment are the poles in question. We have altered magnetic space and more specifically gravitational space. We have reversed gravity which is nothing but an electromagnetic relation due to a varying magnetic density. Charge or voltage potential is nothing but a specific measured need of electrons to release photons (positive pole) and a specific measured need of electrons to absorb photons (negative pole). If photons were to actually pass from one pole to the other the potential would be reduced and the need of electrons would be even.

It is important to understand that small photons travel well in a conductor. Perhaps it is due that their existence can only be maintained in close proximity to an electron and for some reason they do not travel well in free space. Unless we could understand what a single photo-electric monopole is we would only be conjecturing the reasons for easy conductivity of these small photons. When we compare these photons with the traditional electromagnetic spectrum we find that radio waves of the same frequency travel freely in space. This is the biggest reason for establishing a distinction between light and radio frequencies.

Electrons can absorb many small photons and release one larger one such as in the infrared range in terms of heat and even larger photons in terms of the visible spectrum in the form of a spark. The size of a photon is nothing but the number of photo-electric monopoles in orbit. Until the potential is reached in order to radiate larger photons, a positive pole has an increase magnetic density to hold the photons in place. Essentially the oil drop experiment is levitating the drops by evening out the magnetic density. If the voltage potential is sufficient, the drops would rise. If a current was actually passed from one plate to another, the voltage potential would become zero and the drops would fall. There is no current passing in terms of an electric field. What is changed is the magnetic density is greater toward the positive plate than toward the negative, thus reversing the nature of gravity. Instead of determining charge, the experiment alters gravity.

Charge as illustrated earlier is the pumping up of an electron that in turn must increase its magnetic field. In order to exist in a low density magnetic space or with insufficient magnetic flux, the electron is attracted to an area of higher density such as on the surface of matter or closer to the nucleus and in the case of the oil drop experiment—the positive plate or closer to it. An electron is attracted to a positive plate for the same reason it is attracted to the nucleus. The reason is not charge in the mathematical sense but the need of greater magnetic density in order to maintain its size. This is the same as static charge. Outer electrons are so big and exist in magnetic space that it is often not as dense as needed in order to maintain such size that these electrons seek greater density and share such with other electrons. This condition is increased or decreased proportional to the magnetic field produced by the nucleus of an atom. Charge is similar to gravity. The difference is in the nature of every particle and its relationship to magnetic space. Gravity depends upon the polarization of the particle lying perpendicular to the graduated magnetic density in order to provide a net curvature—this would be about half the particle. Charge on the other hand is the particles' inability to sustain the required magnetic flux. Particles need the support of other particles in order to do this. A quantum of charge is based upon the existence of a particle the size of a free electron with the need to share the magnetic flux density of other particles. Magnetic density is greater on the surface of matter.

Faraday established that magnetism could affect photons of light and that there was an underlying relationship between the two phenomena. That relationship is in the spiral helix of light and magnetic space. Faraday's law of induction states that a magnetic field changing in time creates a proportional electromotive force. The opposite of this is that a changing electric current creates a proportional magnetic wave. I do believe that there is no such thing as an electromagnetic wave as conceived by Maxwell. It is only a ripple in the magnetic space.

Maxwell originally intended his models to be an illustration, not an ultimate physical explanation. In the end, he abandoned models altogether. This has been the tendency in science, they simply hide behind and equation without any ability to construct some kind of conceptual model to explain the mathematics and the workings. The danger in an equation, regardless of its truth, is that we still attempt to formulate models in order to explain what we have written in symbolic form. When I discussed $E=mc^2$ while on the subject of conversion, I illustrated that energy does not equal mass otherwise we would simply write $E=m$. E implies total energy of a system and mc^2 defines momentum or the second derivative of mass in motion. Thus energy is equal to mass in motion squared and not just mass as one would say '*energy is convertible to mass*'. Equations can be expressed in concepts and you need to know the mechanics of the model in order to make sense of the mathematics. In Maxwell's final formulation of his field in his treatise (1891, 2:470) he concludes that mechanical modeling was too imprecise to be useful. He said: *'The problem of determining the mechanism required to establish a given species of connexion between the motions of the parts of a system always admits of an infinite number of solutions."* Maxwell went on to say that mechanical models are clumsy with some more complex than others... "all must satisfy the conditions of mechanism in general, that is the conditions entailed by Hamilton's principle (deferential equations of motion), or by the Euler-Lagrange differential equations mentioned by Poincare..."[9]

Models depicted in this thesis are clumsy and geometrically imprecise, but they point in the right direction for others to do a better jog. It is very hard to create a mechanistic model for magnetic space. It is far too liquid or plasma-like in nature. Above all its density varies according to the mass density of the electric field. Space cannot be so applied to electric properties because each field has a localized function either in matter with a moving magnetic field or in a photon with the magnetic field at rest. We may never know what the mechanics are between the two fields. It is probably outside of any form of differential equations because the two fields cannot be separated for analysis. We can only study the electromagnetic relationship and theorize some mysterious cause of the interaction. We might be able to even provide some sort of vortex equations, but this does not give us any idea of the mechanics. We are left to concepts of interaction that we can observe only at a macro distance. We can observe the path of the magnetic field and seem to conclude that it is an infinite number of electric vortexes that appear to be the cause. Beyond this we are left to the mathematics. And since the two fields are inseparable the mathematics is indissoluble. Since all measurements are made at a point in time, usually at the emission and absorption of electromagnetic energy, one cannot conclude that the relationship is the same at every point in-between. You cannot project the equations on into infinity any more than velocity or momentum is going to continue as measured at a point.

In the case of light it was suggested that the light emitter would fire a photon at a velocity relative to the surrounding magnetic density, and then accelerate or decelerate according to various magnetic conditions between the source and the absorber. Equations do not tell us of what will happen in-between the cause and the effect, but reason and clear concepts can. The very same conclusion should apply to radio waves. Light would consider the magnetic field to be at rest and radio waves would be a ripple in the magnetic medium. This is all reasoned to smooth out the Einstein illusion and Maxwellian space. It is not contradictory to mathematics, but it is contradictory to mathematical models that do not satisfy reason. The tendency to produce mathematical models as did Maxwell rather than geometrical models is indicative of a mind wishing to deny the reality of a physical geometry that Faraday so desperately attempted. I think this is a psychological problem with modern cosmologists and relativity physicists. Like Maxwell they think the possibilities are too infinite. But why it is that relativists and cosmologists imagine many universes while the geometric mind sees a much more simple arrangement suggests the psychological nature of modern physics. As mentioned in the beginning—they are Godless philosophical bullies denying the reality of life and death.

Maxwellian space cannot be a simple mathematical convention. It is far too dynamic. It follows a more non-Maxwellian concept of geometrical electric fields of matter existing in variable magnetic space. One might refer to it as magnetic plasma. Some theorists probably come close to this but avoid the obvious in order to establish their importance in being able to see manufactured complexity. It simply justifies their position of importance because their income depends upon it. Simplify it and they might think there is no need of further study. The real problem is that they do not want to face God.

Field theory is the only method in which geometrical understanding can penetrate the mind. The concept of energy in fields as having nothing to do with physical properties is contrary to reason, it fosters paradoxes and makes existence something that Alice in Wonderland would like—"Just what I want it to be." Modern physics only wants to accept that which can be measured and seen but will adopt paradoxes that cannot be reasoned. Is it not far better to resolve paradoxes and consider the unseen?

Samuel Dael

12. The Big Illusion

> Reports of scientific discovery, I'm afraid, tend to give a spurious impression of great progress recently attained. Their thesis is that humankind labored in ignorance for centuries until a few years ago, when the light of wisdom dawned. I think this tendency comes about because discoveries, by their nature, make good stories, while enduring bafflement does not; the storyteller concentrates on what has been learned and ignores what has resisted comprehension.
>
> Timothy Ferris[1]

The Primevil Fireball

It is hard to imagine the very beginning of the Universe. Physical laws as we know would not exist due to the presence of incredibly large amounts of energy, correctly understood as an incredibly large amount of free photons. Some of the photons would combine to become quarks, and then the quarks would form neutrons and protons. Eventually huge numbers of Hydrogen, Helium and Lithium nuclei would be formed. The process of forming all these nuclei is called big bang nucleosynthesis. Theoretical predictions about the amounts and types of elements formed during the big bang have been made and seem to agree with observing the distant universe. Furthermore, the cosmic microwave background, a theoretical prediction about photons left over from the big bang, was discovered in the 1960's and mapped out by a team at Berkeley in the early 1990's. Such conclusions are based upon apparent observation and are the subject of discussion in this chapter.

The perfect cosmology implies that the universe should present a similar aspect when viewed from any point in space or time. If the universe differed in time from its present composition, such as being hotter or denser, galaxies would not form. If gravity were different in time, planetary systems would not exist. Simply put, if the universe was vastly more irregular than at present, conditions would be uncongenial to life. The modern scientist contently accepts that the universe maintains a homogeneous and isotropic condition in space but not in time. Life as we know it would not only be non existent during the scientist's earlier Big Bang, but it will also be non existent in the future frozen expansion. This presents the weakest argument of the Big Bang as far as intuition is concerned. For in this argument the laws of conservation and other universal axioms are laid bare. Such deductions is a universal conclusion without so much as a single effort to first resolve the correct objective nature of light or even truly understand gravity. It is like coming to mystical conclusions without understanding the workings. The Big Bang is almost magical requiring all natural laws to appear upon the scene at an instant of time. Creating the universe from a single moment could just as well provide the same creation elsewhere in the immensity of space. Once a single point in space is labeled the origin of the universe then we must of necessity abolish the isotropic nature of space itself. We could also say there is an overlapping of many universes. If there is one in a billion chances for one universe to explode, then we should have many universes where each would have a different set of momenta. The magically orientated minds delight in this instantaneous creation as it has the same scenario as that of a medieval religious residue. The rationally intuitive do not accept such a magnanimous accident.

The Steady State theory, adopted as an alternate cosmology, provided a more isotropic explanation for both space and time. But due to several observed facts, the Steady State theory was cut asunder and finely overthrown in 1965 by the discovery of the cosmic microwave background radiation. As the universe supposedly expanded, it cooled, much as hot air expands and cools. According to modern theory, the cosmic background radiation comes from a residual vestige of this early primeval era, initiated about 20 billions years ago.

Another principle that lends support to the Big Bang model comes from the observance of more distant galaxies as being younger and more vigorous than nearby galaxies. The vary basis of determining distance is how the cosmologist determines to what degree galaxies are moving away. If there is an error in the measurement of distance then there is an error in determining the age of a galaxy. Keep in mind, the farther the distance the farther back in time the observer sees. This assumes distance can be determined by a point light source. Basic reason would suggest that the greater the magnification required to spread a galaxy across a photographic plate, the greater the distance. This would suggest that on the average, galaxies come in similar sizes. Without knowing the diameter of a galaxy, reason would prevent this method. Magnification of say three galaxies will produce various specs in between. One could never determine if they were very close suns or far more distant galaxies. For this reason cosmologists depend upon the color shift to be the determining factor of distance. Such has been referred to as the red shift. If the color shift determines distance, what then determines a more vigorous galaxy? Is it the pattern or shape? Is it the predominance of specific colors? There are more questions than answers—something the cosmologist fails to clarify.

One of the most interesting concepts suggesting the expansion due to the Big Bang is the Olbers' Paradox. It emerges from the assumptions that as the average brightness of the night sky will diminish in distance according to the inverse square law; the number of stars would also increase by the same law. By doubling the distance, a star would yield only one-quarter its brightness. But by doubling the distance, four times as many stars would shine. By measuring two distant shells, both would be equal in brightness. So by each doubling of distance, the brightness of the sky will double. Yet the night sky appears very dark. Unless the universe continually expands in order to red shift the light out of the visible range, the brightness of the night sky should equal the noonday sun.

The conclusion of a noonday sun assumes that there is no absorption, refraction, or polarizing and dark matter between. It also assumes there is a star at every point in the sky at some infinite distance—something the argument fails to consider. Eventually every large photon is going to be refracted and eventually absorbed by cosmic matter much sooner than infrared light. This is based on the fact that infrared light follows more of a straight line than ultraviolet light because it refracts less at matter's edge. Normal refraction illustrates this. Once refracted, the line of sight has been distorted and no longer represents the original source accurately. What is eventually left is light that is less likely to be refracted. A single refraction will soon produce another creating a scattering effect increasing the probability of being absorbed. Only low frequency infrared would last longer. The whole reason that the sky retains certain shades of ultraviolet blue during the day is due to refraction around atmospheric particles toward the earth of ultraviolet photons. The infrared glances through the atmosphere on to another distant planet or star. This is explained more in terms of a sunset. We are receiving the infrared in more of a straight line of sight from the sun while the blue light refracts toward daylight observers on the western horizon. Writers talk of dust filtering the light. This conclusion is conjecture and cannot be proven. Dust does not allow one color to pass and the others to be absorbed like a molecular constructed gelatin filter. Dust particles are too random in distance and size. I am sure that a laboratory experiment could be constructed to pass a beam of white light through a dust chamber to show that blue light would increase in intensity toward the sides of the chamber. There is so much written jargon assumed without any understanding of the nature of light that we actually propagate lies more than understanding.

The Olbers' paradox reportedly does not account for absorption or refraction. The Paradox assumes that all light travels a straight line only to be absorbed by an observer. It is supposed to be the motion of the distant stars and galaxies that suggest a night sky. Would it not be more rational to assume that all of the light glancing through our atmosphere is going to produce a red shift to some distant observer of our sun? If that light comes close to various particles or the occasional dense atmosphere on the way, one can only imagine that eventually our sun will have a high degree of red shift because the higher wavelengths, as they are called, are refracted by atmospheric particles. It is like viewing a sunset to a minor degree. When we look at a star in the night sky, we must ask how many planets, moons, gases etc did the light come close to and refract out the blue and ultraviolet light leaving us a red shift. Whoever dreamed up the Olbers' Paradox probably likes mathematical riddles and does not think intuitively. Regardless of the theory, the dark night does not predict expansion when considered from a sensible view. But let us assume that refraction does much less to light than suggested and the Doppler Effect is the real cause of the universal red shift.

The red shift is supposed to amount to a loss of energy and while the light from distant galaxies does red shift, it is assumed that a loss of energy is due to blue light becoming red or increasing in wavelength. When the scientist talks about wavelength, it seems assuming because light does not wave in a medium nor does the classical Doppler Effect (according to the Special Theory) make any real sense. The real question is, "What does the scientist really think light is when they use the term waves?" They consider them electromagnetic and they come in colors or frequencies. But how do they know this and how do they measure such small waves? It is basically done by calculation and by special instrumentation that determines the refractive index, but that does not yield geometrical understanding of a wave. What we do understand is that the measurement of light waves is not done, as one would measure a single curve from one crest to another. As visible light is measured by the amount of refraction, broadcast waves are measured by the time it takes an electrical current to move from one end of an antenna to the other end and back again—the longer the antenna the longer the wave length. Measure your TV antenna and consider the velocity of light in a wire and calculate how long it takes light to move that distance. You will get a length per second of time, but that is a frequency that an electric current moves through a wire and does not demonstrate the structure of a light wave.

The mathematics is only a conventional means of talking of waves or frequencies in the case of radio. The conventional geometry in **Figure 12-1** is only a mathematical convention for electromagnetic wave propagation.

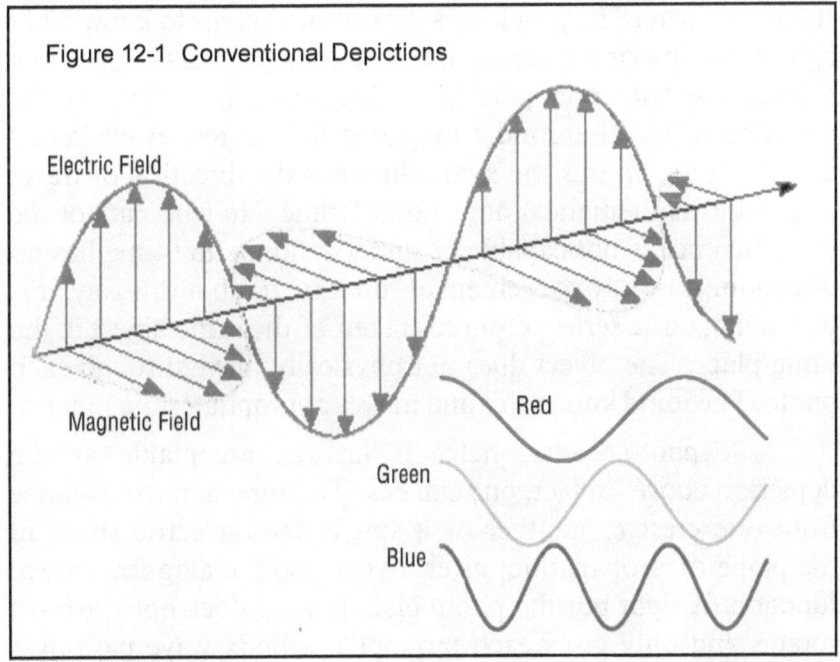

The above is not a geometrical model simulating reality. You can place a list of sales graphically (no numbers) on an Excel form with dates listed vertically and places of transactions listed horizontally and the mind visualizes those sales at particular times and at particular geographical locations. In the above the depictions of light are also a mathematical graph putting time on an *x* horizontal axis and the quantity of each field on a *y* and

The Einstein Illusion

z axis. The mind is forced to treat the graph as reality, because it does not have a geographical model to visualize. Graphs are dangerous when the reader cannot visualize geometrically. The graph then becomes the physical reality rather than a mathematical representation. There is a failure on the part of physics to create a corresponding geometrical visualization equal to a plot on a graph. Light does not move as depicted above, yet writers plaster the mathematical depiction in text books year after year without any new understanding as to what really happens with light as to structure. We are so use to putting time on a horizontal axis because that is how things seem to move that the new observer of the above does not know that the arrows in the above represent reversing currents in and around a stationery wire. The problem is that no one seems to know what light is and no one wants to think of it as particles—especially a photo-electric monopole field that sets out a helix in the direction of travel around a magnetic field at rest as explained in chapter 10. In this the x coordinate is the direction of travel and not a mathematical depiction of time. He who cannot see the difference is not capable of understanding that time has no direction. It is only convenient at times to graph it this way. It is like laying out a series of photos taken of the same object in the same place. The object does not physically move through each photo. The mind knows this and makes appropriate adjustments.

A photo-electric helix is just as acceptable as the depiction above. In fact, one can easily assume that the distance from one crest to another of a single photo-electric spiral in the plane of propagation, gives a very good analog for a wave function of light but the photo-electric field does not wave—it rotates and only gives each monopole a helix wave path over time. By adding additional photo-electric particles in the plane of propagation gives an apparent increase in angular frequency and also a change in color or the ability to refract more to a change in magnetic density due to the larger diameter. **Figure 12-2** does not depict refraction because there is no change in magnetic density from side to side. All density change is linearly as a spiral helix passes through glass.

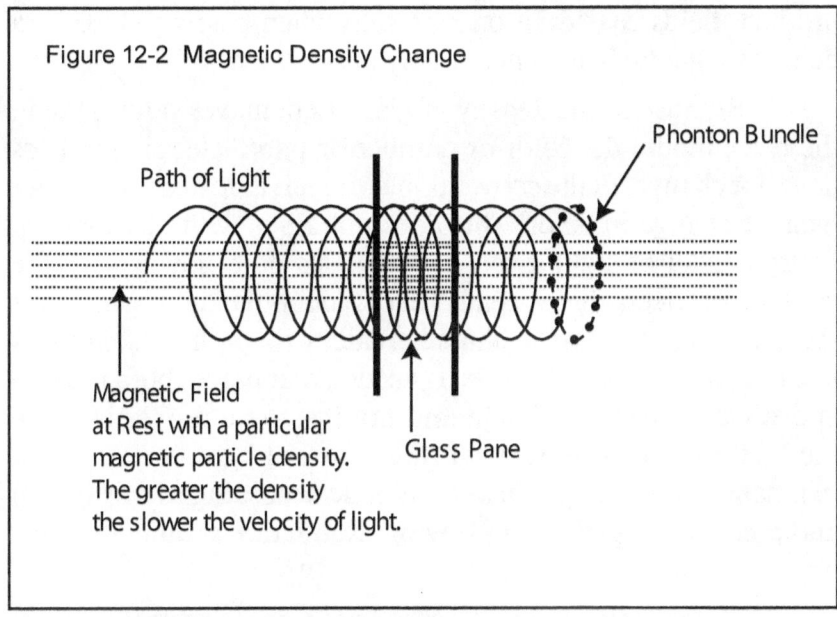

Figure 12-2 Magnetic Density Change

The magnetic field at rest is represented by the small linear dots. For concept purposes only, the path of a photon spirals around a certain magnetic monopole density at rest with a linear velocity equal to the velocity of light in a given magnetic density. Upon entering a dense field, such as a pane of glass, the helical path contracts as it gives up momentum to the higher density magnetic field. Upon exit the momentum is returned as the velocity of light is increased. This model does not fit with solid particle physics and is only a representation of an electric field spiraling around a magnetic particle field at rest. The concept of a light wave fits in terms of the spiral, but the color is not determined by the distance between peaks and troughs, but is determined by how many monopoles are in the plane of propagation. The cause of refraction is due to incident entry and exit and not a right angle entry into a pane of glass. The larger the diameter of the spiral the greater will be the refraction because one magnetic side will be denser than the other upon entering the glass pane at an angle. The greater the number of monopoles the greater the mass and the greater the color frequency or the way photons affect receptors or the electrons in chemicals. Solid particles do not refract unless they produce fields that drag on one side when passing close to a denser magnetic field generated by another mass.

Because of the density of glass, light moves slower, but in the helix model the color or number of photo-electric particles does not change. In the conventional models a slower wave might mean a change in color. This does not agree with observation. Every model of light given in textbooks does not explain the conflict between the Special Theory denouncing a wave in a medium and the particle characteristics of light having mass as it leaves the electron. A model is needed that proves both particle and wave properties of light and not just wave properties. The model should also demonstrate the nature of light slowing down in a dense gravitational field (magnetic according to this thesis) and picking up speed upon leaving. Refraction should also be a byproduct of the same model.

Before considering the spiral helix as the perfect solution we should be reminded that there is a paradox between radio waves and the helical model. The only true argument is really how the photon is created and how it is absorbed. Coming from an electron it is understandable. Also, the helical particle transmission agrees with light as a piece of the electron but obviously does not agree with radio wave propagation. With electrons there is a loss of mass. Does a broadcast antenna loose mass like tungsten filament? Will everyone's household wiring eventually disintegrate? We have a paradox with radio waves. Thus, this concept was dealt with in the previous chapter. A particle or helical model now seems far more plausible.

The classical wave equations of the wave concept are only conventions. They do not predict the geometry of light. The conflict is that the wave mathematics is based upon classical wave phenomena and not particle emission. In as much as there is a rainbow of colors within a beam of light it should indicate the contrariness to the classical wave phenomena. What would happen to several frequencies of the classical wave in a physical medium—they would cancel, interfere or in some cases strengthen other nearby frequencies? Such is not the case with light. Light is fundamentally different in mechanics and geometry even if the mathematics is similar. Light has almost an infinite number of frequencies (more appropriately colors) passing the same vicinity with no interruption—save it be the presence of matter. The filtering of frequencies is also completely different. This in itself suggests particle characteristics more than wave phenomena in the case of light. You can jam radio frequencies because they are magnetic waves. You cannot change light by jamming it with a particular color. Waves and particles do refract but for completely different reasons. A wave refracts because of a change in electric density while particles refract because of a change in magnetic density. With particles it is a change in magnetic density. The idea of a complete electromagnetic spectrum might be a nice concept when you can place every mathematical wave frequency within a complete graphic, but there is one paradox in doing this without any geometrical explanation—you deny the particle characteristics of light. Very long radio waves can penetrate walls, visible waves cannot and yet x-rays can. It is like the middle of the spectrum is limited. If we separate radio waves from the spectrum as being different phenomena we can understand that heavier x-ray photons with many photo-electric teeth can cut through flesh. On the other hand we could say that magnetic waves can pass through almost anything, but dissipate much faster.

If a single visible photon particle as measured in background radiation can travel the immensity of space, why is it that radio waves dissipate and weaken at such short distances? Background radiation is far lower on the so-called electromagnetic spectrum than radio frequency. Background radiation is photon radiation. Radio is magnetic wave propagation. There are so many different frequencies beyond the visible; that what may be thought to be radio may actually be light and vice versa. The essential conclusion is that even radio waves are not like other waves in a medium because the medium in question is not properties of matter, but of the magnetic field. There is a distinct difference. This difference allows you to separate frequencies in both light and radio waves.

But they differ in terms of penetration. Light does not pass through matter because it hits its own substance. With radio waves, matter has only a limiting effect with the higher magnetic density—the ripple still passes because the ripple is not affected by any electric field. It is only affected by the change in magnetic density due to that matter. Radio waves do not have an electric component to bounce off bits of electric matter. Once you take radio waves out of the equation of particle light you do not have the wave properties inherent in the Doppler Effect. You have to explain things differently.

The red shift is not due to a Doppler Effect. Even through the development of spectroscopy led to the surprising beginnings of the expansion theory. The argument is that light emitted by a moving star comes in successive quanta. If the successive quanta are pushed close together by an approaching star, they suggest blue light; successive quanta expanding from a receding star correspond to red light. Light quanta do not squish or contract. If they did it would be between one photon and another and color is not measured between photons. Each photon itself has its own color. It is the motion of the star that may cause a blue or red shift, but not by the Doppler Effect. Given refraction and particle characteristics, questions arise because a quantum of blue is going to remain blue before and after refraction. Light changes direction, but not color. The red quantum is always red before and after refraction. A particle of blue light does not become red, otherwise light will lose its mass along the way just as an electron looses and gains mass. If this is the case, and this will be discussed later, the universe is not expanding, but is relatively static and ever changing. The distance between one blue quanta emitted by an electron and another does not change the color of each. The only way to lose energy is to lose mass and that would be the necessity of losing a chip off the photo-electric spiral of light in the same manner that an electron emits a certain number of mass quanta in order to maintain magnetic equilibrium. So how is it possible for a photon spiral to lose a piece of the action? Remember a reduction in energy is a reduction in...?. Now in terms of the

The Big Illusion | 288

Doppler Effect you have to apply the classical wave concept as found in radio waves. The Doppler Effect works in no other way. The cosmologist tries to argue that the blue photon is still blue, but because of its source's receding velocity the photon appears red like a bullet fired from a receding rifle. If this rout is chosen we have to abandon the Special Relativity principle.

A Doppler Effect is contrary to The Special Relativity principle because the velocity of light is not affected by the motion of the observer or its source. If velocity is not affected, how do you get a Doppler Effect? In earlier chapters one might think that I abandoned the Special Relativity principle. This is a gross assumption. What I did abandon is that the observer did not have a privileged reference frame. A magnetic density was inserted to preserve the observed effects of Special Relativity Principle without countering reason. The process was to remove all paradox. The electromagnetic density of both magnetic and electric properties also solves the particle/wave duality for light. It also should be clear that there is no such thing as a Doppler Shift when it comes to light. If you disagree, let us continue.

In the laboratory the physicist has been able to create rotating mirrors and other apparatus to demonstrate a source of light moving toward in one case and away in another. If we place this apparatus within a relatively consistent magnetic field density on the earth as was done in the Michelson-Morley experiment, and notice that there is a red shift and also a blue shift depending on the motion of the reflectors, an explanation is in order. If light is reflected off a mirror moving away from the observation point and if the light source is stationery relative to the observer, one would have to think there is a loss or gain of momentum or energy. It is difficult not knowing the apparatus. The subject is not detailed sufficiently as was the Michelson-Morley experiment and it becomes an added difficulty in dealing with reflected or polarized light. Keep in mind that polarized light is reflected light and this subject will be dealt with later. If there is a color shift, another explanation is in order. Polarization may or may not be the answer, but it would be appropriate to bring this subject after understanding polarization. If both the rotating mirrors and binary stars are in a relative stationery magnetic density, they could provide the same conclusions, but consider binary stars only in order to remove polarization. How binary stars work requires something not yet considered. In other words, how can a photon gain or relinquish some of its mass depending on whether a source moves away from a magnetic field at rest or towards an observer at rest relative to the magnetic field? I do not think it can is the traditional sense commonly understood.

Keep in mind that electrons emit photons based upon the condition of the magnetic density surrounding the electron. This was illustrated in the helical model and the electron level. One can easily assume that the greater the density the longer it might take for photons to spiral off. Like a charge, the pressure can build up to such to the extent that the light emitted might be more blue given a greater density and red given a lesser density than normal for a given element. The specific element in stars is burning and every atom is receiving (building up charge) and sending out photons (releasing charge). Remember the experiment in atomic clocks that, when moving easterly verses westerly, would experience a more slowing of the clock due to the increase in the apparent magnetic density. Such would be the case of binary stars moving toward us rather than away. The stars are actually moving through a more apparent magnetic density as they approach us and through an apparent weaker density as they recede. This could be due to the magnetic field of each star increasing or decreasing pressure upon the magnetic space surrounding both stars thus affecting the net magnetic field density upon every electron or atom releasing a photon toward the earth. Each direction is going to produce a different result. In one case the atoms delay emitting charge and in the other the charge release is advanced sooner. If the charge builds up, the emission is going to be bluer and if the emission is sooner than normal, the result will be red. The color shift is not visible, but because of the type of element burning there are certain spectral lines that shift—not directly because of motion, but because of magnetic density.

The emission or absorption lines from the burning element in question seem to shift in the spectrum. This cannot be the case for the gases that omit a certain photon are the same, but each photon is larger or smaller depending upon its net charge. The missing color would appear to shift as the net charge shifts, but the emission color on each side of the absorption line is a slightly different color due to the magnetic density. In reality it is not the absorption lines that shift because the atom does not change. It is the color that shifts because of the magnetic density. This color shift will also be illustrated in polarization. Note **Figure 12-3**.

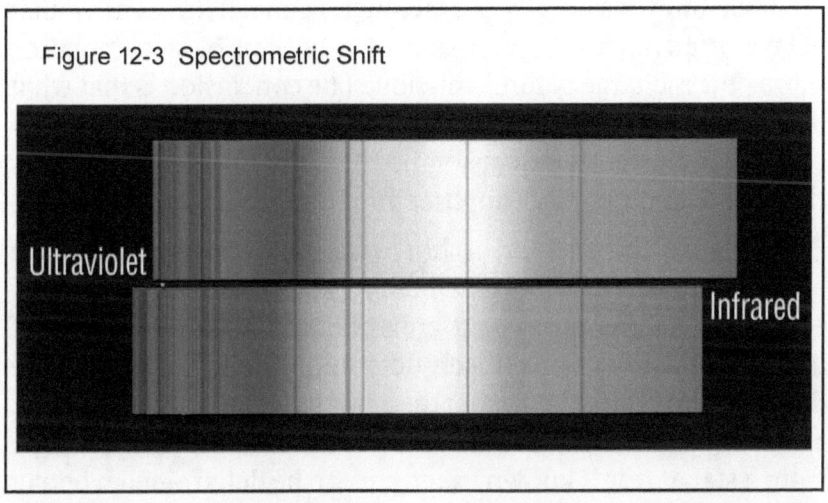

Figure 12-3 Spectrometric Shift

Spectrographic analysis cannot detect any shift in color for what was not visible in one view becomes visible in the color spectrum of another view. Every view would produce the same spectrum of color. This is the same with polarization. It is only when a light source has the absence of a particular color, which is characteristic of the gases of the burning element, that the absorption lines seem to shift. They are not moving. It is really the color. It is clear that every element produces particular emission or absorption lines in a particular pattern peculiar to that burning element. So when the lines shift we know that the spectrum has really done the shifting. If one could see the entire spectrum this would be noted, but we only see a small section of visible light where the invisible becomes visible and the visible becomes invisible producing the same spectrum.

Now why spectroscopy uses the Doppler Effect to demonstrate a red shift is somewhat contradictory to relativity. It requires a medium, yet the medium for light does not exist. In this case the Doppler Effect is not part of the equation. The Doppler Effect would work for a wave or a particle, but not for a system where the electric field spirals around a magnetic field density at relative rest because the velocity of light is not contingent upon the source or the observer, but is relative to the magnetic density. As illustrated with binary stars, what was assumed to be the Doppler Shift was really a difference in magnetic density that delays or advances the emission of a photon by accumulating more or less charge before emission. The conclusion is that what has been demonstrated for binary stars does not necessarily cancel the universal expansion because the magnetic density is the culprit and not any supposed motion.

Since the spectra of hot stars yield lines showing the absorption of hydrogen and helium and since the spectra of cooler stars show prominent emission and absorption lines of calcium and iron, it has been determined that the greater the distance these stars, the greater the absorption lines shift to the red. Because of this, it is assumed that since the expected wavelength from a star at rest is known exactly, it can be determined whether distant stars move away and at what velocity. This conclusion is premature because there are too many paradoxes about what light really is, including the duality of particle and wave. We can assume the Big Bang enthusiasts are correct in the data, but not necessarily in their conclusion. As demonstrated in various examples the magnetic density can have profound effects upon the behavior of light.

Up to this point, explanations of refraction, wave verses particle, and magnetic density variations are spurious but do raise questions that justify further clarification as to what light is. The above theory using binary stars does not explain away the red shift of galaxies or the universal red shift at large. Every means of proving the universal expansion directly or indirectly comes from the red shift theory. Answer the cause of the red shift and the condition of the universe is explained. Answer the cause of the red shift and both the Olbers' Paradox and the resulting background radiation can be explained. In the broader sense, it cannot be magnetic density as illustrated with binary stars, and if the universal expansion is the cause, so let it be. But if by some other means the red shift appears, the universal expansion and the primeval fireball is just a big illusion.

Friction In The Universe

When the physicist draws strange conclusions the average man ponders those conclusions with a sense of bewilderment. Thus far, the quantum particle nature of light, maintaining a constant velocity relative to a magnetic field density has explained the Special Theory of Relativity without the distortion of reality. The same helical model could help to explain the red shift in a variety of ways. Using the concept that light structures an electric field of photo-electric particles spiraling around a magnetic field at rest suffices to show that the smallest unit of light describes the smallest unit of an electric field and also of mass. If more than one photo-electric particle orbits the plain of propagation, the total number would indicate the quantum size or color of the photon. Reduce the number of photo-electric particles in the plain of propagation and a proportional loss of mass of the photon would also equal a color shift to the red. Matter in the universe moves from a higher level of organization to entropy until gravitation eventually crushes matter into a new creation. Might also light as matter mimic the same? A large photon describes a higher level of organization while a single photo-electric monopole manifests total entropy. The question remains, will light always travel until absorption comes upon it or can it also loose a quantum of mass along the way?

Black body radiation may explain this principle. Suppose a sphere about the size of a basketball had one hole. Within this sphere, imagine many reflecting mirrors positioned on the inside surface such that if visible light of a color temperature of 6000 degrees Kelvin can be directed into the hole, little to none would return out of the hole. Like a black hole, the sphere absorbs practically all of the light shot into it. If light does find its way out through the hole, the quantity would be minuscule in comparison to the light entering the hole because of the positioning of the reflective mirrors surrounding the inner surface.

If the sphere absorbed more light than it emitted it would put on mass by the added photon particles. But material body radiates energy not seen and with it, a proportional amount of mass. All in an effort to maintain some sort of equilibrium, the total energy absorbed equals the total energy emitted, mass included. If the body emits and absorbs in perfect equilibrium, it can be called a black body. But the size of the light particles emitted by the sphere comes in greater quantity but smaller in size. It is the same principle of contraction only the magnetic field is not included. This suggests that a large light bundle (high volts) sifts like a clod into its finer constituents. One large photon enters the hole and is absorbed or reflected, but many small (low voltage) quanta exit through the surface. The eye does not see these quanta because the color temperature lies in the infrared range. The hand measures this radiation by placing it near the surface of the sphere. If a black body absorbed all frequencies through a certain hold and yet maintains a temperature of say 300 degrees Kelvin, that same body would only emit quantum colors somewhere between microwave and low infrared size photons. The radiation emitted from a 400 degree Kelvin black body would be in the low infrared. At 950 degrees Kelvin the color temperature would be high enough to affect the retina of the eye. At 6000 degrees Kelvin the radiation peaks at the visible range. The sun's surface lies at this range.

Now the temperature of the night sky's background radiation measures approximately 3 degrees Kelvin. In principle, if the night sky were absolute zero, it would imply the absence of objects in the sky. Since the temperature measures above zero, even in what appears to be empty space, there exists some ponderable matter with an average temperature of 3 degrees Kelvin. Although all bodies may absorb many frequencies, their temperature determines the frequency they emit the most. So the night sky can be nothing more than remote matter emitting many (amps so to speak) of the smallest color of light or larger photon bundles sifting to the smallest quantum of light. Just as the black body sphere does not reach the color temperature of the noonday light sent into it, so also does the telescope receive a 3 degree Kelvin temperature from the black spots in the sky. So far distant, the noonday photons sift into the smallest quanta. If a black body can grind light into smaller quanta, why not the universe?

If the universe continually expands, big bank theory concludes that the radiation of brighter stars with higher temperatures far back in time eventually shift to the red, not by losing mass along the way, but because the wave is stretched out. This supposes a medium and not a mass particle which photons manifest. Either the background radiation becomes the residue of the smallest of matter possible or a wave is stretched out. As to the first, the background radiation does not predict the Big Bang; it only says that the far distant matter in the universe sends out clods of light only to be sifted to a single photo-electric monopole on the order of 3 degrees above absolute zero. The color temperature refers to the peak size of the photons received and not necessarily the original peak size emitted. Cosmologists are caught up in the wave properties of light without a medium. They think light is constant and eternal, but reason and experiments demonstrate this not to be true. Light is vulnerable because a photon bundle has many pieces. It is not a wave of pure energy. Mass cannot be converted to energy. It can, however, be converted to light because light is mass. The cosmologist confuses the substance of light with the energy of motion. Light comes in quantum sizes because its mass elements come in quantum sizes. If energy comes in quanta then it is because the moving mass comes in quantum sizes. Energy cannot exist separate from physical quanta as cosmologists so desperately try to believe.

If the universe expands, why has it taken so long for the background radiation from the primeval fireball to reach the earth? According to the Special Theory of Relativity, the object of emission will not affect the velocity of light. The same applies to the observer. We would have to move away faster than the speed of light for billions of years and finely slowdown in order to catch the light behind us. That does not explain why the individual photon became smaller. If we did slow down, why is the background radiation coming from all directions if the velocity remains constant relative to the observer? The background radiation should only be behind our motion and not in front. You would think the background radiation to be only one spot in the sky if it is the residue of the primeval fireball. If the elements of the earth are part of the explosion, why is the background radiation a year-to-year continuation? It is as if the explosion lasted a very long time. Argue that the universe expands like rising raisin bread where the observer moves away from all other observers at near the speed of light in order to escape the light of the original fireball, then the velocity of light does not move at a constant velocity relative to the observer and with this Special Relativity dies. The Cosmologist is going to have to start all over rather than contradict what they believe. Cosmologists are so fixed upon the observation of spectra that they cannot see the geometry of magnetic space or the quantum of mass. Their mysticism towards energy is their new religion of denial. They believe in superstitious magic of energy more than conservation in physical existence.

Every theory in modern physics eventually contradicts the constant velocity of light. If light travels at a constant velocity relative to a magnetic field density, the red shift would not prove the expansion. If light is constant relative to the magnetic density then we are not moving other than around the sun and through our galaxy. Just as matter or any black body can absorb clod-like photons and sift them as sand, perhaps in some way the universe

The Big Illusion | 296

at large does the same thing. Can another way reduce the size of a photon, other than absorption and emission? Can a portion of a photon be absorbed? And what becomes of a photon if the probability of one single photo-electric monopole is knocked from its plain of propagation? No experiment probably can be devised. If an individual photo-electric monopole could only be removed proportional to the distance traveled, it would most certainly be proportional to the red shift.

If the photon bundle of photo-electric particles can yield up one quantum unit to the space behind in order to maintain momentum passing through a dense cloud, a universe with sandpaper just fine enough will smooth out the big illusion. If the above principle seems even remotely possible, the universal expansion prematurely produces a state of scientific daring and speculation. Since light can be thought of as a bundle of photo-electric monopoles spiraling around a magnetic field in space, the photon must loose mass in discreet units. How this happens only provokes more questions.

Finding the universe in a state of even isotropy after billions of years since a Big Bang becomes less likely than finding a perfectly sharpened pencil balancing perfectly on its point for millions of years. An explosion soon shifts from perfect uniformity. Any deviation from perfection spells disaster for the future. The universe exhibits too much evenness for such a length of time.

> If the general picture of an expanding universe and a Big Bang is correct, we must then confront still more difficult questions. What were conditions like at the time of the Big Bang? What happened before that? Was there a tiny universe, devoid of all matter, and then the matter suddenly created from nothing? How does that happen? In many cultures it is customary to answer that God created the universe out of nothing. But this is mere temporizing. If we wish courageously to pursue the question, we must, of course ask next where God comes from. And if we decide this to be unanswerable, why not save a step and decide that the origin of the universe is an unanswerable question? Or, if we say that God has always existed, why not save a step and conclude that the universe has always existed?[2]

If the universe has always existed as both infinite in space, mass, energy and the ever changing geometry of galaxies, suns, planets, particles and photons; and given enough time, then all light traveling through the far reaches of space will eventually red shift to the lowest color quanta. A static universe with a certain percentage of predictable friction does not contradict the entropy cycle of nature.

Some prefer the change of space, time and mass as their denial of mortality of existence. Others prefer creation as from nothing rather than the immortality of right action. Accept the entropy of all things physical including the size of the photon, and things will change, given infinity to accomplish the task. Why does the theorist marvel at strange theories and why do they bask in big illusions? Perhaps the denial of oblivion hides behind their awe.

The Missing Complementarity

Observation coupled with reason and conservation in existence will maintain the proper dynamics of reality. Observation alone will not do. Without reason and conservation, a model for reality cannot be molded. Customarily, the scientist regards a model of reality only as a suggestion for a particular view. Claiming only to aid thinking, the theorist creates models in an attempt to suggest and not actually describe a particular situation. Coming primarily from the tendency to carry a model beyond the scope of validity, the scientist avoids total commitment.

> There may also arise an unconscious resistance to any changes required by increasing knowledge that cannot be made to fit the model.[3]

The scientist protects a new theory from erosion by calling the accompanying model a symbol of reality and not an actual description. Too often the geometry of objective reality suffers because of this over concern of being wrong. For this reason, the modern scientist too often places little importance upon physical models. Correct or incorrect, the contemporary physicist forgets that the physical model rather than just a mathematical depiction completes the reality process. The mind requires a corresponding geometrical model to the mathematics in order to balance the terms implied in the equation. In other words, the model tests the terms. In a limited sense, the equation only acts as a stagnant verb in the whole scientific process. The terms in conjunction with a physical model yield the dynamics of a complete thought. If the model is incorrect, it is better to change the terms rather than change the meaning of the terms. Science will not do that because they are hiding their strange belief system and want you to believe by frustrating you with a turn of meaning of the terms sacred to you. The importance of a physical model reveals one's thinking and if avoided it indicates fear, denial, and a need for approval. Consider the following:

Subjective	Axiom	Objective
Meaning of the terms	Mathematical equation	Physical model

A model only mimics the terms used in the same way the object receives the action of the subject. If the terms need to be altered that is not a reflection of a bad model. The problem is that most scientists want to hold on to strange terms and avoid a model because it preserves their warped ideas. They would rather talk in universals rather than talk plain. The excuse comes from the idea than complex things cannot be plainly displayed. This is not true. The real problem is that complexity helps the authority to exalt their position and ego rides the horse of pride.

As the mathematics represents the axiom from which the terms and the model pivot, the meaning of the terms must stand the test of an objective model. Strange terms will produce paradoxical models even if the axiom appears correct. If something appears strange in the model, check the meaning of the terms. If the terms express a paradox, change the model. Relativity and modern quantum theory have produced a duality of contradictory models, because the terms contradict.

It becomes arduous in a book of this scope to cover all the accepted models. But, for a good model to stand requires that it exhibit a complimentary view of both classical and modern physics. The helix path of light discussed thus far compliments terms used in both classical and modern thinking. Only one remaining theory in modern physics remains unanswered. It is polarization. But this remains no more difficult than the many contradictory models in modern physics. The most pronounced models of light, currently in use, provide for many paradoxes in both terms and geometry. In one single point, the debate between the wave and the particle model of light still goes on. This should not be if a good model is put forth.

Modern science avoids a solution by accepting a wave-particle duality. The disciplined search for a proportional third model that unifies the two is not considered. This duality, enunciated by Niels Bohr in 1928, was called the "principle of complementary." It states that the scientist cannot simultaneously apply a particle description and a wave model to the nature of light. The use of one model or the other can be used, but never both at the same time. Both the wave model and the particle model agree with observation as long as only one observation is made at a time. In this, the double exposure is avoided. Interestingly, the equations between the two models do not contradict. The greater wavelength in the wave model corresponds to a smaller mass in the particle model. As also a greater frequency in the wave is proportional to the greater mass in the particle model. The two models correspond well in the equation $E=mc^2$ but the meaning of the terms clash with any model attempted thus far so the model has been avoided since relativity. Avoiding a complete complementary model prevents the duality from being two views of an undetermined correct physical model.

Modern physics created a duality out of the complimentary principle by giving up the possibility of a correct model to compliment the two. The complementary principle should have stated that both wave measurements and particle measurements compliment the descriptions of a single undeveloped model. Instead of using the term wave of classical dynamics perhaps the word helix path would explain more about the path of a particle bundle. Neither a particle nor a wave can adequately describe the actual nature of light because these models balance the equations with contradictory terms.

A helix field of photo-electric monopoles can describe both a particle and a wave pattern. The total number of photo-electric particles measures the quantity of mass falling upon a photographic plate and the same number, within a plain of propagation, measures a certain refractive index when passing through a change in left-to-right magnetic field density. In this way, the helix field compliments both the particle and wave model. In addition, the helix hypothesis complements both classical and modern theory. Relativity means only what it says, "Relative to a frame of reference"—the reference being the relative magnetic density rather than the observer. Maxwell's equations work and Faraday's model becomes as real as he desired. Most important, the mathematics of modern quantum theory remain but the laws of probability of finding a real solid-like mass is proportional to the mass of the photo-electric field spiral and the density of the magnetic field in the immediate vicinity and not a probability due to observation.

How then can the current model of polarization and the red shift be explained by the wave and particle duality? It would be well to add a new hypothesis about polarization as an alternate model to the textbook analogy. If this model is accepted, the red shift can also recede in open shame.

Just as refraction can be explained in a different way, can also polarization be clarified? Due to the high variation of magnetic field density within a small slit, light bends upon entering the opening. But with polarization the light never makes it through. Something differs about polarized light in that when the slit aligns in the proper direction, the polarized light passes uninhibited. Using the rope and fence textbook analog, wherein light becomes the rope and the picket fence the polarizing filter, the up and down wave energy put into the rope passes through with no obstruction of energy. But turn the fence horizontally or whip the rope from side to side and the energy of the rope never passes. In some way the electric field of photons stop at the polarizing filter unless the photons align parallel to the slits in the filter. Whatever the structure, it passes the filter only when the filter runs parallel to the reflected surface creating the polarized light. The conclusion is that reflection polarizes light. Filtration only limits what has been reflected.

When the polarizing filter or slits rotate perpendicular to the reflected surface, the newly structured light does not make it through—in some strange way, the structure of light changes upon reflection. What model describes that structural change? The accepted model of an electromagnetic relationship maintains a right angle condition between the electric and magnetic field. Not symbolic, this is a reality to the modern physicist. Just how an up-and-down textbook analog of polarized light passes through when the real relationship of the electric and magnetic fields run perpendicular does seem a bit incomprehensible. The textbook analog of polarized light does not fit the textbook electromagnetic relationship. All photo-electric fields bouncing off a surface with an incidence and reflection of approximately 45 degrees will run horizontal to the surface of reflection. Exactly what happens to the helical model? The same could be asked of any model. I see no explanation that makes any sense.

It must be apparent that polarization needs a better geometrical model than a mathematical wave function. It could be said that the photo-electric field sweeps out a circular helix upon emission and an elliptical helix after reflection. The physicist talks of circular and elliptically polarized light, but the use here generalizes the meaning too much and oversimplifies the geometry. A geometrical model of polarization must explain not only what happens upon reflection, but hopefully help resolve the universal red shift at the same time. Believe it or not, they are related.

A simple experiment can be constructed. The items needed include: a strip of blue, green and red poster card, a photographic light source, a photographic polarizing filter, two white reflectors and a photographic spot meter. Lay the strips of colored card upon a table in the same order as in the visible color spectrum. Put blue to the left, green in the center and red to the right. You can reverse it and the results will be the same. Make sure the colors resemble the pure colors and not the high intensity combinations such as cyan, yellow and magenta. Make sure they are not dark colors. Place on each side of the table a white reflector—a kind used to diffuse and soften the light in portrait photography. Behind the table, place a photoflood lamp pointing down at a 45-degree angle toward the colored cards with a spread sufficient to hit the two white reflectors pointing down also at a 45 degree angle to the color cards. From a camera's point of view, point a spot meter down at the blue card. The effect should be the same as looking down at a pool of water with the sun coming from the black sky. The reflected glare dominates the diffused light coming from the side sky, or in the case of the experiment, the reflectors. By correctly placing a polarizing filter over the spot meter, the light intensity drops. Either the light intensity coming more directly drops if the filter runs perpendicular to the table, or the light coming from the side reflectors drops if the filter runs parallel to the table. In other words, the polarizing filter cuts the light, either the more direct light or the defused reflecting light. As might be assumed, according to the spot meter and the eye, the more direct light measures a greater intensity.

More light does come from the light source, somewhat as a direct glare, than through the reflectors. But sight the viewfinder upon the red card and the reverse is true. Keeping the polarizing filter perpendicular to the table still cuts the more direct glare coming from the light, but the meter reads the intensity of light as greater coming via the reflectors. A contradiction seems apparent. By sighting the viewfinder upon the green card, it will be noticed that the meter intensity remains somewhat the same regardless of the filter rotation or the path of light being selected. Can an answer be found? The facts are that the blue card reflects more direct light than diffused light. The red card is the opposite in that it reflects more diffused light than direct light. Finely the green card is about the same for both. The spot meter measures total intensity and not color.

There are two things that reflection seems to indicate. The first is that the photons realign parallel to the table surface after impact and second, the photons shift in color to the red. Traditional wave analogs suggest that all waves are reflected parallel to the table, but do not suggest a red shift. If a red shift does occur, one needs to explain how high frequency light changes to low frequency light and still conserves momentum. It would be like one blue wave becoming two red waves to explain the situation.

The hypothesis continues: Assume that a single photon bundle spirals around magnetic space and breaks up into two photons upon impact on the blue card in such a way that the remaining photons spiral in a furrow parallel to the table. If the slits in a filter run parallel to the table, the eye would receive the full intensity of the polarized light (both photons). If the filter were to turn perpendicular to the table, the filter would prevent some of the furrowed photons from passing or would refract around the slit to the sides of the camera before hitting the film. Consider the light, coming from the source via the white reflectors and then reflecting off the colored cards. Would this not produce a breaking up twice? The first reflection would yield two photon bundles perpendicular to the reflector and the second reflection would yield a furrow of four photon bundles perpendicular to the side of a molecular peak on the card surface that was facing the meter. The only light coming from the reflectors that can reach the observer would reflect off the sides of those peaks, thus keeping the furrowed arrangement almost perpendicular to the table. Except for the scattered light from other parts of the room, the majority of the light reflected from the colored cards will thus lay furrowed perpendicular coming from the sides of many molecular peaks or it will lie furrowed horizontally, coming more directly off the tops of each molecular peak on the card's surface. Depending upon the rotation of the filter, either the more direct light or the twice-reflected light drops, but not both. But why is there a difference between the blue card and the red?

Consider that each card reflects a particular color and absorbs the others. The reflectors however reflect all colors. If there is a color shift this means that ultraviolet light becomes blue, blue become green, green becomes red and finely red becomes infrared. With diffused light the shift would be doubled because of double reflection. It should also be understood that visible light peeks in the blue, green and red as far as intensity is concerned and drops off in the ultra violet and infrared. The glare intensity from each card would be due to saturation in which all colors are reflected. The reason the blue card has more intensity from the lamp directly is because the blue card is actually reflecting ultraviolet light that has been split to become more visible in the blue range. There is not enough intensity available in the twice-reflected ultraviolet light to be visible in the blue. What intensity there is will be reflected by the green card. This condition would be less in the case of the green card because there is plenty of blue to become green from the more direct light and also ultraviolet light to become green. With the red card, the conditions are more intense. The more direct red light has shifted to the infrared out of the visible spectrum, but the double reflected blue would now be red. The reason that there is no reduction in intensity of the twice-reflected light from the red card is because there is more blue and green light becoming red than ultraviolet becoming blue or green. If this analogy is correct, the green card would have about the same intensity of glare as with the diffused and the red card would produce additional glare from the diffused light due to plenty of intensity from the blue. There are fewer green photons than blue in white light so the direct light intensity is less. A good spot meter notes this fact. If the photographer is shooting a dominance of red, he must open his aperture more than if blue. There are actual color marks indicated for proper adjustment.

The solution comes by understanding that, regardless of the light path, the cards reflect more of one color than the other two. Although they reflect all colors due to saturation, a peak in one color results from the reflection and absorption properties of the card in question. The above concepts can be explained in **Figure 12-4** showing two curves.

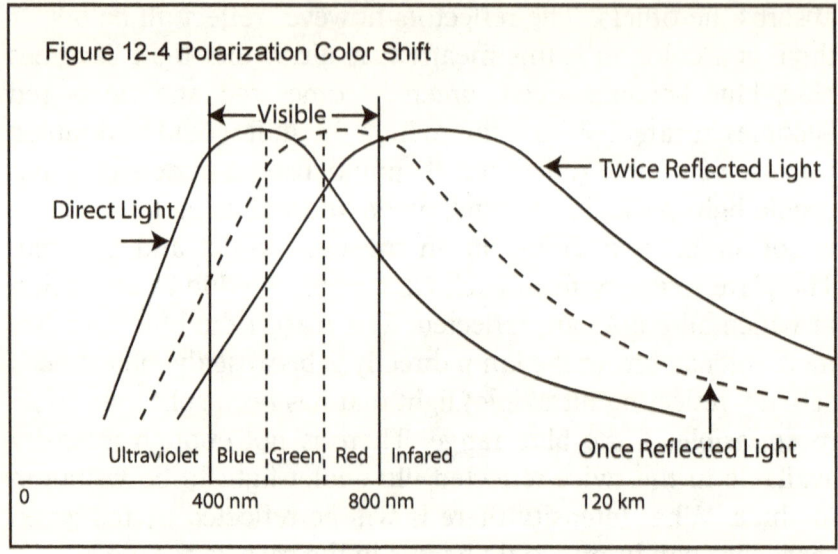

Proper photographic light peaks in the visible spectrum, a range that includes blue, green and red. The intensity is far less in the ultraviolet and infrared range. All three colors equal white visible light. Notice, however, that it begins to drop in the red. This curve is direct light without reflection or as illustrated, non-polarized light. The second curve to the right represents light that has been reflected twice, assuming that 800 nm is double that of 400 nm. The graph refers to wave length in terms of nanometers. If blue light is about 400 nm (a specific size photon that refracts twice that of 800 nm yielding half the wave length), then the second curve represents reflected light as each photon is split in half. The second curve is extended because the scale increases in length the further right. This is also due to the scale that begins to extend rapidly the further right. It is extended to focus on visible light and extended to condense very long wave lengths. The intensity remains equal assuming no filtering. The only difference is the color shift. Note the increase in red of twice reflected light and the drastic reduction in blue light. This would explain why red light remains relatively constant. You could draw a dotted curve between the two for once reflected light and the red would remain about the same but the blue would be stronger in once reflected light than twice reflected light.

Although a polarizing filter cuts about half the light intensity, most is probably polarized but not all. If a polarized photon furrow arrives aligned to the slits in the filter, the intensity would pile up as illustrated by the appropriate curve. When the intensity is greater from reflected light we use a polarizing filter to cut the intensity by running the lines perpendicular to the reflected furrows thus cutting about half of the reflected photons. Without the filter the intensity would be as high as non-filtered light but would shift to the red. As can be illustrated, reflections set the stage for the potentiality of the red shift. Aligned filtration simply reduces the intensity of reflected light. What the eye sees after filtration appears as the intensity of the first curve superimposed on a portion of the second. Both curves crisscross the visible spectrum and develop only a loss in intensity, partly due to some of the loss of photons and partly due to a color shift out of the visible spectrum. The apparent color remains for the most part the same, but the very small red shift results from the fact that there are not enough ultraviolet photons to fill the void. Essentially when light originates as ultra violet and becomes green, blue becomes red and what started as green becomes infrared. The loss in one frequency appears as a gain in another. But since the natural peak lies in the blue to green spectrum, very little ultra violet light appears to fill the void created by this loss. Thus reflection or splitting can cause a slight shift to the red.

The two curves demonstrate why the red card offers more glare when filtered while the blue card has greater glare unfiltered. The green card has about the same intensity filtered as it does unfiltered.

It seems appalling to use the concept of polarizing filter as a measure of color shift. But we are not talking of filtration or refraction but rather reflection. If electrons can absorb large photons and emit smaller ones, it makes perfectly good sense to suggest that light breaks up upon reflection. The majority of the time, light is refracted in directions other than the line of sight. If light is not refracted, it is broken up by reflection. This was the case with the black body thought experiment. Filtration seems to cause a color shift, but this is not the case. Filtration does not shift to the red as will reflection. Filtration simply omits the intensity unless it is a colored filter. Some textbooks suggest that the sunset is due to dust particles filtering out the blue and green. This suggests that the writer did not understand the nature of light. The ultraviolet, blue are refracted around dust particles while the red curves only slightly with the green and yellow in between. The color of the atmosphere particles has little to do with the red intensity other than the temperature of those particles will enhance the refraction or curvature of light. The atmospheric particles do not filter the blue. They refract it. If a sunset had a position of origin, those standing under it would see blue light coming down—as if looking up in the sky. At a more distant place they would see more yellow sunset with a possible green halo at the bottom. Usually it is too light to capture such a transition. Refraction is why we have a blue sky even though some textbooks call it filtration by hydrogen and oxygen particles. A rainbow is not filtration either; it is refraction through droplets of water, just as the water droplets are not filters neither are dust particles. Refraction can be around a dust particle just as it is through a water droplet. This is all due to the variable magnetic density next to the particle and within the surface of the droplet.

What about the early morning sky? If you are high enough and look out the window of an airplane, a very early morning sunrise will have a red halo. It will not be pronounced as a sunset because of the atmospheric temperature. The red shoots into outer space while the yellow and blue refract to the lower elevation of the observer. This is not filtration. It is again refraction.

In returning to reflection, we can consider the Olbers' Paradox to be a complete misnomer. Consider a mirrored elevator. There is a sudden stop and the lights go out. For a split second you see red reflections of the other travelers. This is the continually reflected light that breaks up until we can no longer see the infrared photons that continue to divide and divide. Should not the light continue to bounce from one mirror to the other? Certainly the eye and the clothing do not absorb the blue, the green, and the red too abruptly. Reflection and refraction to a degree accomplish the same thing. It eventually moves all light to the invisible infrared range—reflection breaks up and refraction changes direction more for blue than red.

Just because reflection may cause light to shift to the red does not disprove the expansion, but it does illustrate the vulnerability of light in that an individual photon can break apart so easily as demonstrated in black body radiation. If the immensity of space can be considered to possess a measurable and relatively consistent amount of matter with an occasional ability to remove a photo-electric particle from the plane of propagation you will eventually have a red shift. This is not the old notion of tired light as might be construed from a wave of light in a medium, but relates to actual physical properties of light known to possess quantum particles. When you are talking about infinity of time you cannot dismiss the possibility. Even if the odds are one in a trillion, that is sufficient to eventually shift all light to the red or to a single photo-electric quantum of 3 degrees above absolute zero. Consider **Figure 12-5**.

This is similar in concept to light passing through a pane of glass. The images are completely out of proportion in order to illustrate the concept that when a photon enters a dense magnetic field it slows down and yields up some of its linear energy to the magnetic system it is passing through, but upon exit there may not be sufficient conditions to yield up the momentum needed for the photon to accelerate. In this case, the photon spiral might leave directly behind an individual quantum unit of background radiation that might continue even if the larger photon is refracted. It could very well be that each photo-electric spiral does not contain all of the quanta in the same plain. They

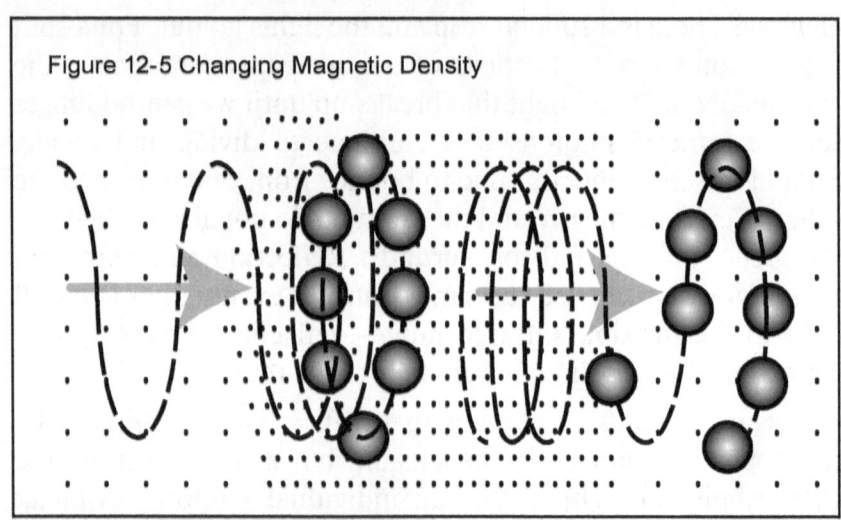

Figure 12-5 Changing Magnetic Density

might tend to arrange themselves in the path of the helix rather than simply create an apparent path. Perhaps some experiments could be done with a specific thickness and density of glass by sending a certain light with emission lines through it. If the lines shift after passing through glass, the theory is demonstrated.

If all light shifts to the infrared naturally by reflection, polarization, and also by refraction, the emission and absorption lines would shift as well. The spectrum will always appear normal because what was invisible will shift onto the visible spectrum and what was visible will shift off the spectrum. Without this normal color shift we would not experience darkness. Light slides down the red hill until it hits bottom. But the linear motion of a single photo-electric monopole will eventually be converted to angular motion within the atom when it enters the proper magnetic field density within matter. Enough magnetic conditions will eventually collect many single photo-electric particles into a larger particle. At several stages linear momentum is converted stationery angular momentum. At an infinitesimal point, the sun shines.

Total entropy does not exist in the universe. Ask yourself how much light of any size does the sun collect? You can use the same argument used in the Olbers' Paradox. If there is eventually a star in every direction then some light from every star that is not refracted, reflected or absorbed will fall on our sun. Add to that more refracted or reflected light coming from other directions, just how much light is absorbed by the the sun? It must be a tremendous amount. Just like the earth that may absorb certain larger photons and eventually sift them to a very low form of radiation, the sun having a very high magnetic density fluctuation will sift light out in the visible range. If every star, planet or moon in this sense is a black body where it absorbs equal to what it radiates, then matter is constantly being created through absorption and radiated at many levels. We think of the sun as burning up its fuel, but at the same time it might just be creating more. We are so susceptible to observation that we cannot consider the power of the unseen. The sun is probably not in perfect equilibrium, but the point is made.

Unlike a cold reflective surface the sun does not reflect light as does the moon. The sun probably chews it up like a black hole before it spits it out again. In this sense there is no such thing as a black hole that does not emit much of what it takes in. Just because Black Holes suck in all of the visible light does not mean that they do not radiate very low voltage photons. Stephen Hawking used key scientific works that have included theorems regarding gravitational singularities in the framework of general relativity, and the theoretical prediction that black holes should emit radiation, which is today known as Hawking radiation (or sometimes as Bekenstein–Hawking radiation).[4]

A definition of a black hole is a region of space from which nothing, not even light, can escape. It is the result of the deformation of spacetime caused by a very compact mass. Around a black hole there is an undetectable surface which marks the point of no return, called an event horizon. It is called "black" because it absorbs all the light that hits it, reflecting nothing, just like a perfect black body in thermodynamics. Quantum mechanics predicts that black holes do emit radiation like a black body with a finite temperature. This temperature decreases with the mass of the black hole, making it unlikely to observe this radiation for black holes of stellar mass.

In one sense the theory states that black holes absorb all light, but the theory also states that they emit radiation. This is contradictory unless you are caught up into strange relativity concepts. This contrariness is especially true when black holes are considered comparable to thermodynamics and black body radiation mentioned earlier. The biggest problem with black holes is that everything is simulated and cannot be proven. Above that is the common assumption that gravity has pulling power and is not a variable magnetic density. If it was not for the closeness of the sun's reflection, the moon would be black in the sky. It does not emit visible light of its own, but it does radiate light in the very low range—all bodies do. There is no way not to assume that the background radiation is really a compilation of infinite number of black bodies in the sky radiating 3 degrees above absolute zero—black holes if you prefer.

The fundamental problem with black holes is that it takes gravity equations to infinity, but if all energy comes in discreet quantum units then gravity must diminish to zero at some point because you cannot divide the smallest unity of energy. The infinity of gravity also speculates upon an eventual collapse of the universe. On the other hand black bodies of perfect equilibrium as suggested have no more proof than the Olbers' Paradox, but it illustrates that the universe is very dynamic and always changing within certain laws. For a universe to blow apart says that light and matter must have behaved contrary to physical law at some beginning. This defies reason and destroys the meaning of conservation. One's god is then a god of magic and not responsibility. When a child grows up and becomes a physicist does he really put away childish things? There is nothing wrong with understanding mystery, but understanding means to make sense and not further one's enchantment.

In every case when relativity predicts something there is a paradox that is ignored. We are more fascinated by a bug-eyed-monster-universe than we are of understanding. The universe is dynamic and it is not a machine gun that runs out of bullets. To proselyte the universal death of the big bang will not mask the scientist's fear of oblivion. It becomes obvious that they are nothing but bullies playing games with our minds. If they can prove the improbability of God surviving an exploding or imploding universe, they have demonstrated the ultimate denial of the existence of God. Only by accepting entropy on a cyclic basis and understanding that observation of itself cannot create a reality, will the principle of complimentary between field and particle yield the reason every man so desperately needs. Are we really being responsible in trying to understand the universe or are we playing alluring games of fascination?

13. A Matter of Intelligence

When the heart spills its life upon the uneven ground, listen for that which glistens in the air. Death is only an honest trial. It is the light that overcomes.

Samuel Dael

The Denial of God

Analytical thinking often leads us into endlessness steps of system counting void of a good sense of human understanding. On the other hand we have analogical thinking that keeps our focus on the relational picture where we see that some things can be seen more fully through conservation and harmony. It is like having wisdom and faith in human nature rather than fear for what the ignorant might due. Because we see such things as inequality through analysis we tend to force fairness and do not realize that we eventually bring destruction because we cannot see the relational aspects neglected and the ultimate value of individual responsibility. John Taylor Gatto perhaps understood this when it came to education.

> Our problem in understanding forced schooling stems from an inconvenient fact: that what wrong it does from a human perspective is right from a systems perspective.[1]

Whether in education, family, any business, religion, government system, or even science, analysis often becomes more important than the average human perspective. The human perspective missing is analogical thinking. System analysis does not have broad principles such as allowing free choice, patience, longsuffering and gentle kindness to work naturally. These principles are intuitive, analogical and provide a rudder that keeps analysis in the proper direction. A belief in God does this very well as long as human principles are among one's belief.

Analysis sees only opposition, good and evil, justice and mercy without the wisdom of harmony. Harmony is not compromise or equilibrium in the sense of averaging things out. Harmony is looking for a reason that the electron does not fall into the nucleus. It looks for a reason that gravity works and it looks for a reason when observation is contrary to conservation. Harmony does not accept paradox as true because it is contrary to analogical thinking. Analytical thinking will easily lead one into a black hole. Equations often lead into strangeness because what might be effective at short distances does not apply to infinity. The reason for this is that the equation is based upon short term analysis due to observation rather than the mind's eye making sense of what is seen. Magnetic density was created out of analogical thinking and the need to explain why things fall rather than just accept the analysis of observation. The *why* is the key to understanding and so often we discourage the many questions of *why* by using the remark, "That is the way it is." Asking, "Why" is essentially a different type or application of intelligence? It does not easily agree with what is authoritative. It has to understand for itself.

The atheist by intelligent processes is analytical. Although the theist is often analogical they often become analytical in terms of forcing righteousness. They will create systematic moral laws in order to control bad behavior rather than teach values and personal responsibility through the process of analogy rather than analysis. The Master was prone to use analogy and its parable sister allegory while his enemy always used analysis. Such is the case of wisdom that transcends reason by looking for that intuitive guiding star as individual intelligence travels through life. You can demand justice by analysis or you can work justice by sacrifice. It is giving up one's life unto death for a higher principle that overcomes one's tendency to force a process upon another. When we accept death rather than deny it, we begin to overcome. When we accept God rather than deny

Him we begin to see the light that glistens in the air. When we deny death and God and look for a bug-eyed-monster universe, we build an analytical system around us in order to keep us from facing the inevitable.

Modern cosmologists have built an analytical system around them in order to scare away the reality of life and death. They accept eventual oblivion of the universe in order to claim superiority to it. They deny God in order to rise above him. Their extensive analytical knowledge is used to fashion a maze of darkness so that others cannot perceive their innermost psyche and see their naked ignorance of logical sensibility. Analytically 1+1=2, but analogically it equals three—not in terms of math, but in terms of geometrical dimensions such that two eyes yield the third dimension and two ears yield the third harmonic. Combine the inner eye of spacial harmony and we can determine unconscious motive—something that cannot be seen or heard. The number of relationships in determining unconscious motive is vast—something far more than 2+2. Analogical thinking does not totally rely on the mathematical equation until understood. Analogical thinking is more definitive and qualitative in nature rather than simply quantitative.

Analytical thinking can often make good sound evil and evil become something good. What lies behind the analytical process of turning things upside-down is one's psychological motive. It is almost a condition of spiritual darkness or intelligent wretchedness. It is when intelligence fails to evolve properly. It is an individual problem and has nothing to do with physical reality such as genetics or environment. It is something vastly different than anything related to the electric universe.

Unless the analogical process overcomes our tendency for analytical pebble counting, the ocean of truth will remain in darkness. Particularization from step to step binds the thinker into a linear process and eventually into the awe of oblivion. The analyst, out of desperation, curve-fits reality into a form of denial. Analytical thinking by itself is a no-win process and can only be avoided by forming proper generalizations upon all the steps at hand before counting. Newton may have fathomed this relationship when he said:

> To myself I seem to have been only like a boy playing on the seashore, and diverting myself in now and then finding a smoother pebble or a prettier shell than ordinary, whilst the great ocean of truth lay all undiscovered before me.[2]

Finding the smoothest pebble upon the seashore creates a step-by-step analytic process into endless meaning. When the vast universe seems undiscovered, analogical and intuitive generalizations about pebbles in particular are the philosopher's gold. Instead, modern cosmology thinks linearly and when infinity is reached all of reality is curve-fitted into a personal or psychological denial of oblivion. Of what value is it to particularize the particular? Pebble analysis should retain some essence of intuition. Science seems so sure and probably does not ascertain intuitive generalities as it should. It is the nature of all to generalize the particular and suggest why something might exist a certain way. Proper intuition can aid in making proper analytical steps. Improper generalizations will cause us to curve-fit reality to an unconscious need rather than fathom the ocean of truth. Making generalizations incorrectly is not an intuitive leap, but the satiation of denial. Intuition at every step of the way is the only way to work through a maze of information. Intuitive leaps are circular; they work downward or inward upon the step at hand. Intuitive leaps will magnify while denial simply curve-fits reality into one's personal existence formula. Proper intuition upon particular pebbles can qualify each step. Analysis without intuition does not boost understanding—it can only predict an outcome by extending a few steps into infinity. Analysis does not qualify—it quantifies. Intuition does a far better job in reaching into the past and into the future. When we use analysis to look into the past and the future we tend to distort reality into psychological motive. Intuition does not burden itself with psychological scarecrows. If we neglect intuition we fabricate strange conclusions. It is similar to neglecting the right side of the brain in order to force an analytical conclusion. In his book *"Naked Optimist"* Keith Kelsch expresses the ability to stand and say no actually forces a more clarified yes. Even if an objection lacks analytical support, the fact that the objection is made in what may appear a negative salutation, shows that understanding is not satisfied. What seems positive and optimistic is nothing but counterfeit discourse lacking true understanding. A false optimist defends a paradox by changing the meaning of reality in order to intimidate the intuitive members of society with their vast knowledge and ability to regurgitate mathematical platitudes.

The vast danger of analysis comes when curve-fitting the parts become more important than intuition in maintaining basic principles. Claudius Ptolemy (second century) did this very thing by curve-fitting reality to the analysis of observation. In the Ptolemaic system, the Earth remained fixed in position and all the other planets revolved around the earth. To explain the observed retrograde movement of Mars, Ptolemy changed the nature of space and time by adding a second eccentric motion to the already basic circular motion. Ptolemy should have intuitively re-positioned the Earth in the universe. Intuition certainly could have developed other alternatives. Analysis is masculine-like, but intuition is more feminine. One should not overpower the other at any step. When God created a help meet it was not a servant. Religion jumps to a common conclusion out of prejudice. Like religion science places more emphasis on the masculine rather than the feminine. Curve fitting is a byproduct of subjugating intuition to reason.

Dr. Susan Hyatt gives a definition in her book *In the Spirit We're Equal* such that the Hebrew *ezer kenegdo* for the word 'help meet' meant 'one who is the same as the other.' This would imply that intuition should have equal play in our intelligent processes. Analysis is not subservient to logical thinking nor should analogical thinking be subservient to analysis. Both analytic and analogical methods must work together in equal harmony. As mentioned in the development of meaning and epistemology, it was illustrated that *one* means equal and if *flesh* is a biblical word for 'temporal' then *one flesh* means equality in temporal means. Servitude should never have been a physical requirement. It should always remain a personal sense of equal responsibility. Analytics should never have completely overcome science as it has. History has proven that darkness will follow. Columbus did not rise out of the darkness of the times—he rediscovered what was taught in ancient history. The flat earth mentality was a fall from ancient analogical thinking of the Egyptian and Sumerian cultures. The flat earthers were a breed of analytical thinkers raised on controlling the populous and avoiding intuitive thought. Cosmologists have done the very same thing with modern physics. We are in another age of darkness because of man's analytical denial of God rather than the intuitive nature of trying to know and understand Him.

Curve fitting accompanies the Special Theory of Relativity by redefining space and time. Reality is changed to fit observed fact rather than letting intuition explain the observation when analysis cannot. Many a great mind fail to comprehend that intuition must work with all forms of reason and observed fact should only be considered an active window into reality and not the conclusion. The biggest folly by some relativity writers was to call modern relativity an intuitive leap. This turns intuition upside down. Intuition is not a science of paradox; rather it is a science of making sense where no sense is found with what is observed.

Intuitive thinking develops naturally when the mind's eye preserves the wisdom of antiquity and learns to set aside fanciful traditions. Take the intuitive ideas about God. These perceptions are essential when covering vast concepts—just as certain as ideas about God are fundamental to daily life. They are ideas that express a responsible God that obeys the conservation laws imbedded in the universe and not a God who manipulates the world for magic and for personal need. The ideas about God are not much different than what we expect of ourselves. If we expect something for nothing we are apt to believe in a magical God that has no human traits. If we set up ourselves to gain control we are apt to believe in no God at all. If we seek for honor and great achievement we are apt to think God has chosen us. If we do not believe in a God of human honesty we are apt to make ourselves God through our achievements, hoping our works will be carved in stone. The grandest lie against humanity is to use traditional ideas about God to get power and gain over others. This is what sets tradition into forms that will not reveal a God of human character such as was demonstrated in Socrates and Christ.

The essential problem with God is our death and the accompanying fear of it. Death needles our immortality formulas so we create and curve-fit the world to scare away this reality. Death reminds us of oblivion, decay and a loss of identity. For this reason we would be apt to think that Socrates and Christ were fools. When a principle has more value than life itself we are especially honest in this regard. We face death with human honesty and not with psychological denial.

When we do not attribute the unknown to be intuitively resolved and thus follow bizarre solutions we are apt to curve-fit reality and justify the paradox of strangeness just to prove to ourselves that we are not afraid. There is this psychological attempt in science to deny the responsibility toward honest knowledge in favor of a psychological need to kill God. The use of observation as the only source is only an excuse. The priests and kings during the time of Columbus suggested that observation be the only criteria. We cannot count or observe infinity, but we can intuitively understand it. Looking into the night sky or into the smallest components of matter was once a look into the mysteries of God. Since modern science has matured, it has attracted so many atheists that the Big Bang has now become the justification of atheism over that of evolution. The psychology of both comes by way of the same motive and the excessive use of analytic thinking.

Atheism and the facing of oblivion is the fanaticism of denial. Atheism is usually not an ideology one will espouse against the backdrop of death, but in the case of denial it is important to show your daring in order to find immortality in discovery left behind your eternal demise. It is none other than becoming the bully of knowledge to scare away your own anxiety about death. It is not unlike a daredevil trying to prove his worth by running into death. In reality he wants to see how close he can come to death and live. A soldier will give up his life to save another. This is not denial. It is faith and human honesty for principles rather than the glory of denial. There is a fine line between a hero of responsibility and the so-called hero of denial. It is the difference between honesty and a psychological lie. Such is the problem between atheism and theism which fosters human honesty and intuition. Theism that fosters control is the same as Atheism in terms of psychological motive.

This same psychology is not unlike the ancient idol worshiper who looked upon God as a master magician rather than a figure representing human honesty in the universe. On one hand, if a belief in God satisfies our greed we miss the mark. On the other hand, if we choose no God, we justify our ignorance and our lust for freedom from responsibility. God is the representation of those conservation laws imbedded in the universe. To deny God is to deny conservation and responsibility. The paradoxes of relativity serve this purpose well to those who want to prove their importance without being individually responsible. They use relativity as an escape to look death in the face just so they can feel alive.

Intuitive honesty is the search for general principle and does not come to a conclusion so easily through a step-by-step analysis. If analysis comes first, the thinker curve–fits God and reality into a fabricated mold in order to give support to a specific interpretation. Analysis is a lot like mathematics. It predicts, but does not explain the workings. Mathematics allows us to change the meaning of the terms without upsetting the calculation, but in doing so we are changing the workings and denying any intuition about the terms used. Intuition is that good sense about the meaning of terms. It is that classical square of reason rather than a new conclusion fabricated by a perfect analysis using improper term meanings in order to cloud the geometrical workings of what is really going on.

By requiring an intuitive analog the analectic thinker might first ask, "Did God will the smallest basic particles into being, or did this intelligent mover organize the already existing particles according to universal law already imbedded in existence?" Analysis cannot answer this question, but intuition can. Choose the first and analysis will prove that God is a God of magic. Analysis in religion often chooses magic over responsibility. Choose the latter and intuition will accept that creation organizes the basic elements. Intuition needs to be a part of analysis in order to come to an effective conclusion. God does not will things into existence. In other words, analysis will prove different things, depending upon the intuitive analog. Having the wrong beginning view will only produce the wrong conclusion. If science would avoid curve fitting in order to support their analysis and rework the intuitive analog, perfection of knowledge is far more eminent.

If every rationalist uses the right side of the brain, they should come to the conclusion that intelligence does not arbitrarily make up conservation laws. Intelligence becomes intelligence because it interprets and comes to understand the laws already in existence. Likewise, the conservation of space, time and mass comes from a base analog. Warped space-time is curve fitting. Intelligent thinking does not need to distort the terms just because they are afraid. Of course, no relativist will admit he or she is afraid. That is the nature of denial. They are using the darkness of relativity to prove that they are not afraid and they also intimidate others for not believing as they. It is much like the atheist wanting to force his view upon others. The use of the name Einstein does not convert a sound mind into curvature of reality. The strong mind will object. Einstein himself objected to the direction relativity was going, but could not alter the direction because he did not see the godless motivation in his contemporaries.

If something observed does not make sense, the scientist does not need to redo reality. Simply look deeper into the individual parts of reality for an acceptable geometrical analog.

Einstein attempted this very thing by looking for a unified field theory before his death. By then the rest of science was on its way to magic and the mystery of big bang cosmology. To the best of my knowledge, we have lost any attempt for a geometrical unified electro-magnetic field theory.

Without intuitive knowledge, the rationalist depends upon the pebbles of observation and curve–fits reality in order to end the pebble counting rather than take an intuitive leap. Too much analysis will eventually develop a denial of objectivity and the very existence of matter and the reality of particle field theory of the electric and magnetic relationship. The denial of anything intuitive is the methodology of the atheist and much of science. Human intelligence, which we might refer to as Godly intelligence, remains an intuitive part of reasoning. Analysis, on the other hand asks, "Why would a God of infinite love create such pain, poverty and horror in the universe?" The problem with this analysis is that God is not a God of magic—something assumed by the atheist in his disappointed youth. We assist the lower animals only as they behave civilly. Why should we not expect God to assist man if he learns to behave civilly? Just because man is intelligent does not justify his greed. Something for nothing seems to be the norm far more than responsibility. Intuition answers, "If God remains an objective part of the universe, subject to every law, the answer seems clear. The willingness of God to correct every wrong must be subject to the laws imbedded in existence." God must surely speak, "Except you become one and move with the laws of the universe, the abundance of my Spirit shall be withheld. I do not throw seed to the wind only to be dried up." God obeys law. He is not a God of magic. Unless we obey the same laws in respect to each other, God will not help us. We are left to kick against the pricks and suffer the control of others that we might learn obedience to the laws of the universe rather than obedience to the arm of flesh and that of a controlling authority.

Einstein seemed to like the idea that God obeys the laws of the universe when he said:

> The man who is thoroughly convinced of the universal operation of the law of causation cannot for a moment entertain the idea of a being who interferes in the course of events—provided, of course, that he takes the hypothesis of causality really seriously.[3]

In order to be subject to the laws of the universe, God would need to be a part of existence. This is an intuitive concept, but most analyzing thinkers, including Einstein, give the impression of a belief in an intangible and impersonal God.[4] If God on one hand depends on the laws of physics, yet remains a non-objective part, the universe itself becomes God. Einstein called this a cosmic religion.[5] Einstein's impersonal God did not organize the universe; God became the synonym for the universe. This same scenario has developed over the centuries in Eastern Philosophy. It comes from the mind not understanding the allegory of the eastern sage. It comes from thinking that God is within rather than Godly *action* is within us. The God within us closes off responsibility in the name of outward meekness that simply covers pride while Godly action embraces accountability with no thought of how we appear. Existence requires action to be a part of the total universe and without a supreme intelligence to act and to change within the universe, God simply does not exist as anything distinguishable from the universe itself. Still, in a contradictive way, Einstein said:

> I want to know how God created this world...I want to know His thoughts, the rest are details.[6]

This is like looking for intuitive knowledge before the analysis begins. Einstein walked a narrow line between a personal God and a universal synonym for God. Most who rebel against the traditions of religious magic yet desire social acceptability will follow the same path. Juggling concepts about God is like finding acceptability. The total denial of God is designed to simply irritate social authority. The need of significance surfaces when one disagrees and needs to show a new argument through pebble analysis. The denial of God is just this. But those who do not want to deny God, but resent current authority find other ways. Einstein resented his early schooling when the instructor used fear, force and other methods of artificial authority. The denial of authority precipitates every young scientist to produce scientific publications to impress the controlling structure. Authority itself fuels the fire by continually demanding new analysis.

> For an academic career puts a young man into a kind of embarrassing position by requiring him to produce scientific publications in impressive quantity – a seduction into superficiality which only strong characters are able to withstand.[7]

Einstein challenged authority by setting aside the objectivity of reality. The Special Relativity Principle expressed a denial of established authority by curve-fitting analysis. Einstein's analysis was far clearer than his contemporaries, but his intuition lacked a solid foundation. Analogical and intuitive thinking usually does not get the attention. It is too plain and simple—too sensible. Mystery is far more tantalizing than sage thinking. Those who supported relativity in the early days and those who fought it both failed to explain it. They failed to design geometrical and intuitive analogs upon the first principles; instead, they dot-counted existence and then began to redefine a new universe.

Einstein was a very perceptive analyzer and an intuitive thinker. He abhorred the detail of mathematics and revealed true analogical ability in the photo-electric effect. The General Theory of Relativity was also well balanced between analysis and analog, but it continued with assumptions based upon Special Relativity. If Einstein had used an intuitive analog for relativity as he did in the photoelectric effect, he would not have to put into his general relativity equations a cosmological constant in order to make reality fit his personal concept of God. He would have replaced the Special Relativity Principle with a sound unified theory. Einstein wanted good sense, but relativity and placing the observer at the center of the universe was now killing the God of his youth. Perhaps part of him wanted this and perchance part did not.

Inside the heart of Einstein was a true analog. It just never came out. Once General Relativity was analyzed, the

The Einstein Illusion

prognosis predicted two things in which Einstein objected. Here lies the true Einstein, and if I might say, the greatest. General Relativity predicted an assumed origin of the universe. The curve fitting it predicted, he objected to; but he never renounced the curve fitting of the Special Theory of Relativity. The general theory eventually developed into a subjective interpretation of modern quantum theory and was also associated with the expanding universe. The subjective interpretation of modern quantum theory says that every particle waves as pure energy until the observer sees the particle. To Einstein, his real analog was that the universe exhibits objective, determinate and predictable conditions, but to the quantum theorist the universe became totally unpredictable and observation dependent. Einstein's personal God began to surface. A world controlled by the observer seemed unacceptable. It implied chance. In his objections, Einstein resorted to his now classical statement about God when he said, "God does not play dice with the world."[8] The General Theory of Relativity again challenged Einstein's God. The expanding universe, that Einstein objected to, evolved from a special, observer dependant relativity of Maxwellian space. The general theory eventually endorsed the Big Bang. So horrified by this conclusion, Einstein introduced into his field equations a cosmological constant. This constant possessed the equivalence of a repulsion force to counteract the universal expansion. Troubled by the idea of a universe that blows up, Einstein wrote in a letter,

> "This circumstance (of an expanding universe) irritates me," and in another letter about the expanding universe, "To admit such possibilities seems senseless."[9]

Einstein reluctantly repented of his cosmological constant. He never accepted an indeterminate universe and he only tolerated the universal expansion. Einstein applied the rest of his energy to solving the unified field theory. He continued this direction until his death. As a youth he made a mistake and changed the world. He could not change his mind about Special Relativity but he never even took it to the level in which others that followed did. Einstein did not wish to put the observer at the center of the universe, but the Special Theory did just this. That has become the problem with modern physics even today. At first Einstein did not see the implications of the General Theory, the equations were right and the cosmological constant was only added to keep his image of God. What was wrong was the concepts of light were playing havoc with the General Theory. Einstein did not need a cosmological constant. He only needed to insist the particle field nature of light for which the photoelectric effect should have suggested. The Special Theory was a rebellious dream based upon Maxwellian space. No one would question the equations of Maxwell, but they were now playing havoc with Einstein's greatest achievement. Relativity created a subjective monster and Einstein failed to produce an analog for light that would unify a field theory and solve the ever-perplexing action at a distance for gravity. Gravity was still thought of as something infinite rather than eventually diminishing to nothing as the graduated magnetic density eventually is even at large distances from the solar system. The proper analog of accepting the photo-electric and magnetic monopoles as two differing elements of objective reality could have shed light upon the universal expansion. It also could have brought into focus a unified field theory. Also the philosopher's of the time should have corrected relativity's epistemology. Was it Einstein who coined the expression "warped space" or was it those who followed? The use of the "forth dimension," to Einstein, only implied a mathematical convention, not a reality.

Perhaps the reason Einstein became intolerant of an expanding universe, stirred up the idea that all the matter and energy in the present universe would have originated from a concentrated extremely high density cosmic egg, fireball, titanic explosion or more simply put, the Big Bang. Perhaps it affected his analog of God. The remnants of this cosmic fireball appeared to be detected in what the scientist calls background radiation. But for the most part, the red shift expansion concludes the Big Bang theory. A little expansion here or there, may not trouble one at first glance but the primeval fireball does carry plenty of horror. Theologians of the metaphysical, generally delight in proofs of the Big Bang. It's fantastic and incomprehensible. Einstein and other scientists rejected it. The Steady State Theory preserved one's personal God while the expansion theory turned one's God into oblivion. Many tried desperately to find an alternative explanation, but failed.

The epistemological difference between the theologians and determinists lies in the concept that the universe had a beginning. The mystics say God created the matter first and then the world. The determinist prefers the eternal existence of matter and energy and thus God is not the creator of it. When Einstein came to New York in 1921 a rabbi sent him a telegram asking, "Do you believe in God?" and Einstein replied, "I believe in Spinoza's God, who reveals himself in the orderly harmony of what exists."[10]

Now when any astronomer or physicist talks about God, his colleagues assume him to be either mad or going over the hill. Does the philosopher scientist admit their true feelings or do they shroud the objective belief of God into universal dogma acceptable to all? Both the child and some adults will believe in a personal objective God without too much prompting. If the intelligent still harbor this child-like concept, they will reject the horror of the Big Bang. If the intelligent reject God altogether, the molting fireball becomes their denial of God and of death. As the atheist and his calculating theologian counterpart reside in common ground, they both deny a personal objective God void of real substance.

> God is subtle, but not malicious," said Einstein. This faith, which a physicist needs in order to believe he can understand the universe, is also needed by a mathematician trying to understand his mental universe of number and form. Perhaps this is what Dieudonne means when he calls realism "convenient." It is more than convenient; it is indispensable.[11]

The realist or objectivist position remains the position, which most mathematicians and physicists secretly prefer to take. But when pressed to answer to a Godly belief, each scientist conceals his or her belief in universals. Einstein was no exception.

Opposition to the subjectivist quantum field and the ideology of big bang physics revealed that Einstein's God still played a part in the action of the universe. But Einstein did fail to understand that the Special Theory of Relativity opposed an objectivist solution about God. A distinction between God as the sum total of the physical universe (Einstein's universal response) and a personal objective God lies in the concept of intelligence as separate from objective reality. If intelligence and physical existence originate from the same physical existence, then God does not exist in terms of intelligence or physical substance but he exists only in terms of law—that which he did not create but is. If intelligence and physical properties are of a separate universe, then man is more than the physical. If God is only intelligent and not physical, then God is less than man.

The physical, biological and psychological workings of humankind structure a system more complex than that of the universe itself. Does God have less than this? God should be more complex than man and not less. God too must have physical properties of some kind. Perhaps man is totally entropic and God not. This would imply something more than man. God must at least be as complicated as the workings of man if not more. If all things are physical only, then all things human will end in the future expansion or the collapse of a black hole. If God is not physical as we are, perhaps believers of magic consider that God snapped his fingers (excuse the physical pun) and the universe blew up. It makes more sense to reject this conclusion and accept the eternal and separate nature of intelligence and also the eternal physical and intelligent nature of God.

Now if God exists both subjectively as intelligence and in some way also objective in from, the Great Organizer would not have emerged from the same primeval fireball nor would the intelligence of man have evolved from with a big bang. For this reason the subjectivist's God having no objectivity and the atheist's denial of any objective God, together they enjoy the same Big Bang cosmology—a denial of death and also of God. The reason for this is that a primeval fireball destroys everything objective.

Probability of God

Epistemology requires at least nine reference points in order to define reality limiting the ability to alter terms. Terms can be shifted around at will unless they are nailed securely by one of these nine points. They are vertically subjective, predicative or objective and horizontally as masculine, harmonic and feminine. The harmonic is nailed clearly understanding that it is not a compromise, but the superimposing of two views into one. It is exactly how the brain works in observation using two eyes. The harmonic is three dimensional. This concept has resided in the background since the beginning of this thesis. Just as reality has subjective attributes along with the predicative and objective, so also does God. Just as views are masculine and feminine they are brought together in harmony. The intelligence of God and that of humankind must bring reason and intuition into a harmonic wisdom. The Spirit of God and of humankind must struggle with justice and mercy in order to establish the beauty of good works. The objectivity of God and of humankind has existence and that implies intelligent responsibility and creation according to conservation. Only through patience and longsuffering will the perfect and harmonic society arise. Without it there is nothing but a *Platonic Idiom*—an illusion in today's philosophy.

Just as some treat God as objective only or spirit only or intelligence only, so also do they treat reality? You can measure the foundation of the philosophy of a man by the completeness of his God. Philosophy has failed because the meaning of God follows a limiting psychological profile in each and every person. The various attributes about God have been debated since the beginning of philosophical inquiry and still seem to be debated openly in such low-level college courses as literature, psychology, science, and philosophy. Usually due to the instructor, the challenges to the existence of God develop from personal misgivings about existence itself. Throughout this thesis, the occasional relations about God have been intentional. The hypothesis I hope is clear: that one's image of God has a direct relationship to the realities he or she accepts or rejects, is laid bare. Proof cannot be given as to the existence of God any more than proof can be given of the particular existence of an inhabitable planet somewhere distant from our own. Science only knows the probability of such things and not the actual existence. Considering reality as a whole, God's existence can be probable, but never non-existent. To some, that level of probability is very high—to others, very low to nil. Believe it or not, the degree of probability comes more from intuitive understanding than rational analysis. The psychological ramifications emerge from the basic denial of human fear of oblivion, death and rejection. If it was not for these psychological problems man would have an easier time coming to some probable sense. For this reason it is essential that at least the atheist consider the principle of probability, just to make sure his or her psychological fear is not the controlling factor.

Just as the laws of mathematical probability can conclude that given a sufficient count, the statistician will find one being more intelligent than another upon the earth, continue this probability count and one reaches the most intelligent being in a given galaxy of the universe. Include another galaxy and the probability count will simply continue until we have the most intelligent of all. The average individual has no difficulty accepting this probability process, but to some professors, the universe could not be old enough to give rise to a civilization more intelligent than our own. They argue that since there's no particular reason why there should be any other intelligent beings out there, additional civilizations would just be wasteful thinking. If the universe had no beginning as the polarizing red shift would suggest, the one civilization argument reveals a grandiose ego sealed by the unrecognized fear of infinity which translates into a denial of God. Carl Sagan was so eloquent in his denial before his death that he spent a great deal of effort establishing the Big Bang. It is the same with every scientist because their upbringing suggested a god of magic and control and not a god of responsibility. If the child prays for something that he does not get, he concludes there is no god. Later when these minds consider the poverty and death in the world they also conclude there is no God. Death still haunts them but they deny it by accepting or formulating philosophies that actually kill God and make Him impossible to face.

To accept a Supreme Intelligence above all others within a particular galaxy of the universe, bothers no scientist, but to call that intelligence God may or may not. Concluding that God created the universe as from nothing bothers not only the faithful; it also bothers one's intuition about God. By placing God above and beyond comprehension seems only to bury reality and man's individual obligation to abide the laws of the universe. Nothing sinister appears by giving God the capacity of organizing the elements, even the planets. But for God to break the laws of conservation by making something out of nothing smashes the scientific, the intelligent and the measurable. The creation of the universe as if from a puff and a bang describes the same product, as do the religious medieval magicians. True creation is organization and wise judgment about the elements and altering them over long periods of patience through natural law. It is certainly not a magical process.

The Spirit of God has been called "Right Action." Under this predicative definition of God's existence, the scientist appears not to be at all bothered by crediting one being as more upright than another in a given order. An intelligent being may move in perfect Godly Action. But, to the superstitious and to the metaphysical believer, the Spirit of God defines some sort of physical incomprehensible such-ness and not right living. The psychological power of the witch doctor prevents one from understanding that "right action" resides in an objective being in order to move correctly as an integral part of existence. Right action is not separate from being or becoming. Also, right action in God is not separate from His being.

And last, the fear of death, pain, and suffering control the beliefs about the objectivity of God's existence. Religious prophets do not make it clear as to the many references about God. Whether they refer to the two physical entities, observable electric-like or unobservable magnetic-like substance, or whether they refer to Godly action in man, all this seems difficult to interpret without a three-reality foundation. Christian prophets separate the three more distinctly than others: The Father, being the Supreme Intelligence, the Holy Ghost being the Spirit of Right Action and Jesus Christ being the resurrected physical model. The oneness or unity of the three comes from equality of intelligent action and not equality of substance.

The Spirit better refers to the right action of objectivity (with a capital (S)pirit) or to some eternal spirit-like a magnetic substance beyond the grave (as in spirit). For example, the Spirit of Love refers to action, while, "...for a (spirit) hath not flesh and bones and ye see me have" (Luke 24:39) refers to substance beyond the grave. The above distinction provides the most challenging corollary about the objective existence of God. Action cannot be action and intelligence cannot be intelligence without the ability of substance to act. If God is not also objective, how can he act? Both scientist and layman avoid this conclusion. They feel that this makes God an anthropomorphic being. The Jews of antiquity had no difficulty with this concept. Whether the objective existence of God resolves from magnetic particles, photo-electric particles, a yet an undiscovered third or a combination of any is not meaningful at this point. But the probable existence of God implies position or location without trying to name the place. "For God to be small enough to dwell in the heart," does not refer to location. This refers to action as that of "heartfelt action" or a feeling in the heart to do something good.

Just how one interprets a religious or philosophical writing will tell a lot about one's reality formula. Religious writings, meant to be nebulous and allegorical, allow the average individual a chance to curve fit the inherent teachings to personal needs based on underlining fear. Due to this curve fitting, the teachings remain more palatable until the student rises out of fear and trembling. From that time fourth, the student mimics the sage and speaks in paradoxes for the benefit of only those who have ears to hear.

Becoming too clear about reality may cause some to oppose and destroy. But the foundation of modern physics and the vast knowledge that the future holds hangs in the balance. A third part reality is long overdue. Listening to the poetic message may seem more palatable than understanding, but philosophy and science must maintain proper interpretation to lift the truth out of mysticism. This also applies to the prophet. The spirit of interpretation lies in the ability to keep the realities separate. Only to the fearful do the realities thus combined mix up and shift from side to side that the blind see but see not and the deaf hear but hear not. All too often, the learned still remain terrified. So woe unto the prophet or scientist that still fears, for when the blind lead the blind, both fall into the ditch.

All things then have subjective, active and objective existence. Heaven, intelligence, man and God exist in a three-part reality. I would like to say that Hell is truly psychological unless intelligence was required to physically start over as suggested by some aspects of Eastern Philosophy. He who might understand will. Nonetheless, to deny sound meaning will warp reality. Modern physics has mixed up the terms and seems to foster an upside-down model of reality. Relativity likewise seemed to be molded to fit a God of magic and the medieval desire of something for nothing.

This same miss identity of reality had a similar impact on the other sciences, in particular the theory of evolution laid down by Charles Darwin. The biological sciences carry the same distortions of reality. For example, the measured age of organic properties through nature's atomic clocks assumes equality of magnetic conditions since the beginning. Another point about the self perpetuating ego of biological evolution professes a principle of anti-entropy while all of modern physics maintains total entropy as a universal law. Can evolution be placed in proper perspective? If so, I trust that someday I might be able to write "The Darwinian Descent." If it takes as long as this book I am afraid it will never be written. The key is not in the contradiction of science but in better meaning and better analogs of understanding. For science, psychology, philosophy and religion must become one. They must move with the Tao of existence and not against it just to satisfy one's psychological propensities.

The Tao Of Physics

The duality of particle and wave fostered a quantum concept wherein observation created objective reality. Under this duality, the particle does not exist until observed. Until that time, light was said to exist as a wave. This quantum leap led to a unification of modern physics and Eastern mysticism. Once the duality of particle and wave can be resolved, the distorted aspects of mysticism can be explained under a more appropriate classical religious philosophy.

"The Tao of Physics" by Fritjof Capra, creates parallels between the worldviews of physicists and mystics. This has been hinted at before, but never thoroughly explored. The quotations in this section come from Capra's book and demonstrate that the same quotations can illustrate an opposing, but classical view illustrated throughout this thesis. I will use Capra's own words to show a classical view without paradox and with more responsibility. Those who develop strangeness are running away from responsibility and take the masses with them.

The tenets of Hinduism, Buddhism and Taoism do show striking parallels with the latest modern experiments are due more to the inadequacy of a single complimentary physical model. The parallels superficially stimulate more than reveal a direct correlation. The compilation of strange models in physics prevents a complementary model from arriving on the scene. A harmonizing model would solve the paradoxes and dualities.

> Thus the aphorism of Einstein, "As far as the laws of mathematics refer to reality, they are not certain; and as far as they are certain, they do not refer to reality.' Physicists know that their methods of analysis and logical reasoning can never explain the whole realm of natural phenomena at once, and so they single out a certain group of phenomena

and try to build a model to describe this group. In doing so, they neglect other phenomena, and the model will therefore not give a complete description of the real situation.[12]

Full of paradoxical situations, physics carries suggestive images without precision. The lack of a complete model therefore provides strange conclusions based on observational cause. The philosophy in Eastern mysticism has been taken for granted because the paradoxes in modern physics sound like the paradoxes in mysticism. It should have been understood that mysticism describes the nature of a philosophical allegory and should not be taken at face value. The same applies to classical Judo-Christian religion, which should also not be taken literally.

At the beginning of this thesis, the word "mysterious" was more appropriately applied to great religious philosophies. Mysticism was treated as obscure thought or speculation while the mysterious implied a lack of understanding. The great error in mysticism, or mystery if you will, surfaces when the paradoxes are taken literally. Mysticism was designed in an effort to cloak the truth in allegory.

> Indian mysticism and Hinduism, in particular, clothes its statements in the form of myths, using metaphors and symbols, poetic images, similes, and allegories.[13]

Just as Western thought has its roots in Judo-Christian mysticism, so too does oriental philosophy find root in Indian mysticism? Mysticism should be avoided all together as a branch of metaphysics and more properly laid down as a spiritual psychology and as a philosophical technique in understanding that which is avoided by man because it requires responsibility. Metaphysics means beyond physics, but too many think of it as a physical reality beyond what is seen. This too is fine, but the tendency to think of magic seems to negate its value. Religion, mysticism and even metaphysics is an attempt to describe the psychology or action of responsibility in one swoop while physical science describes cause and affect in mathematical points in space and time. One view is stagnant while the other is active.

Western theorists misunderstand oriental philosophy when they miss the mark and talk of oneness between the observer and the object. This oneness treated literally or physically, produces superstition and magic. Seen only by the wise, the true meaning of oneness lies buried in allegory. Interpreting the allegory as magic or subjective only depends on the degree of one's fear of reality. The errors come by mixing the psychological or spiritual with the physical. True interpretation is predicate based and not the comparing of subjectivity to objectivity as does one compare physics with Taoism.

By equating observation as subjective or giving the objective the magic of mysticism alters the intent of the mystic. The oneness used in comparing mysticism with relativity becomes a mixture of reality and not the psychological attitude or action of the whole intended. Switching reality from the objective found in physics to the subjective found in mysticism turns reality upside down. This swinging of reality from the objective to the observational and eventually to the extreme subjective side caused Einstein great difficulty. He wrote in his autobiography:

> All my attempts to adapt the theoretical foundation of physics to this (new type of) knowledge failed completely. It was as if the ground had been pulled out from under one, with no firm foundation to be seen anywhere, upon which one could have built.[14]

Not a simple misunderstanding, the new mind in physics blatantly refers reality to observational cause. The verb relationship in reality came to be described as objective existence instead of action of existence. Just as Einstein seemed troubled, so also the simple-minded wrestled with the magic in the doctrines of relativity. Once too often, they interpret mysticism as they do the mathematics of relativity. They do not understand the mysteries of the sages.

Mysticism implies a paradox, but underneath one should be able to find the psychological (action) that will remove the paradox. Western and Eastern mysticism reveal the emotional psychology needed to solve the allegory and the paradox.

> The Chinese mind was not given to abstract logical thinking and developed a language which is very different...Many of its words could be used as nouns, adjectives, or verbs, and their sequence was determined not so much by grammatical rules as by the emotional content of the sentence.[15]

While mathematical equations do not have emotional content, mysticism stimulates a certain psychological reaction by stating a paradox. Only when a certain level of understanding can be reached will the paradox be solved. The following expresses a few examples:

> Whenever you want to retain anything, you should admit in it something of its opposite: Be bent, and you will remain straight. Be vacant, and you will remain full. Be worn, and you will remain new.[16]

These paradoxes do not describe physical reality anymore than the psychologist's statement to the protective parent, "In order to keep the child, you must learn to let the child go." The understanding parent sees no paradox. But the fearful parent takes the paradox literally. The allegory or paradox lies before us as a stumbling block in which to overcome the fear of reality and not to define it as has been done by the modern physicist.

Eastern philosophy does not have a monopoly on metaphors and similes. Judo-Christian philosophy has its counter parts. What greater paradox can be found than in the statement, "He that findeth his life shall lose it" (Matt. 10:39). The understanding mind sees no paradox. They see that when one is only concerned about finding himself will he become lost. Designed to compare the physical with the spiritual or the subjective with the objective, the religious or mystical paradox does not equate. It compares in an effort to stimulate psychological insight and responsible action. It has nothing to do with the physical finding or loosing something. Thus comparing the Tao with modern physics is a complete destruction of a true objective solution.

A professor of Eastern Philosophy placed the following on the board, "He who dies before he dies does not die before he dies." Before he gave his interpretation I sat and pondered

A Matter of Intelligence | 338

it. I quickly came up with a Judo Christian answer and said it to myself in the following way in order to remove the paradox: "He who dies *spiritually* before he dies *physically*, dies before he dies *because he is already dead when he really dies.*" The eastern interpretations would be "He who dies *by giving up becoming and seeks the being inside, dies* before he dies, *and therefore* does not die when he dies *because he is not dead.*" The two interpretations are very similar and illustrate the nature of paradox in allegory. Both Western and Eastern mysticism stress the active, but many interpret the Eastern and subjective and not active. This is the problem with many who cling to Eastern Philosophy. It becomes a subjective escape rather than an active responsibility.

To understand the deeper meaning of life, the individual must be at one with his or her environment. Not an objective blending, for it involves not a psychological mental process, but an active process. It involves moving with natural existence and not against it. The actions of the Taoist sage, or any religious philosopher, thus arise out of intuitive wisdom spontaneously and in harmony with natural law. One does not force the mind or the heart, but merely adapts one's actions to the movements of the Tao or the whole. This has nothing to do with the physical properties of quantum light.

> In the words of Huai Nan Tzu, "Those who follow the natural order flow in the current of the Tao". Such a way of acting is called wu-wei in Taoist philosophy; a term which means literally "non action," and which Joseph Needham translates as "refraining from activity contrary to nature, "Justifying this interpretation with a quotation from the Chulowang—tzu: "Nonaction does not mean doing nothing and keeping silent. Let everything be allowed to do what it naturally does so that its nature will be satisfied."[17]

Nonaction means no action toward things that are not natural, but does not mean nonaction towards things natural? This type of paradox can be misleading and is often misapplied. No action means that we do not resist moving with what we naturally should. Like a young tree, becoming one with that which is natural means oneness or agreement the natural wind. Wisdom bends with the wind, but never alters its purpose. To be at one with the universe implies appropriate action and not inappropriate action. The wise bend with existence whether it is social, legal, psychological or political. They render unto Caesar that which is Caesar's and unto God that which is God's. Whatever those in authority say, we observe and do; but those in authority "do not after their works: for they say, and do not" (Matt. 22:21 & 23:3). They wait for the stillness and then speak the truth in paradoxes. He who has ears to hear will hear.

Equality and oneness do not unify in the sense that the young tree becomes the wind. Equality and oneness unify only in the sense that the tree moves with the wind. The individual participates in the universe. Thought, observation and existence express parts of this active whole. Observation itself arises from the movement and agreement with the universe. Observation particularizes the action of the whole and does not describe a conversion from one part of reality to another. The mystic requires us to see that apparent reality and interconnects the whole rather than alters it.

Existence can only be objective and action the relative motion of existence. Never convertible, objectivity and action cannot exist without the other. Understanding this non-conversion will reveal the true meaning of not only Eastern mysticism but it will also reveal a better model for quantum physics. Instead of the participator observing objective reality into existence, or instead of the relativist curving space, the mathematics, like a mystic expression, may simply show that a variable magnetic field density will generate all the illusion in physics by altering the direction and velocity of light. We think the observer is in control, but he is only a participant seeing as he moves with the universe. Seeing does not create as the popular interpretation of the Tao implies. Seeing is moving with the Tao.

Relativity does not move with existence. It creates its own by changing the meaning of space and field. Relativity treats space as objective and field as subjective or a mystical action without substance. You cannot separate matter from the field any more than you can separate energy from matter. Matter is not space. Matter and space define different realities. But to the relativist, the fields associated with particles do not exist as real dimensional substance. Fields exist to the relativist as probability waves of pure mathematical quantities void of objectivity.

The origin of this relativistic reality switching grew out of Mach's philosophy. Although Einstein rejected much of the 20th century outcomes, Einstein did lay the foundation by acceptance of the premise of Mach's philosophy.

> According to the physicist and philosopher Ernst Mach, the inertia of a material object—the object's resistance against being accelerated—is not an intrinsic property of matter, but a measure of its interrelation with all the rest of the universe. In Mach's view, matter has inertia only because there is other matter in the universe. When a body rotates, its inertia produces centrifugal forces.... but these forces appear only because the body rotates "relative to the fixed stars," as Mach has put it. If those fixed stars were suddenly to disappear, the inertia and the centrifugal forces of the rotating body would disappear with them.[18]

When a body rotates, its inertia produces a force against the magnetic particles in space and not against the fixed stars as Mach implied. If those fixed stars were suddenly to disappear, a decrease in associated magnetic density would result as a partial vacuum in place of each star, but the inertial forces of the remaining rotating bodies still remain. The forces could only be minimized relative to the surrounding magnetically filled space. Gravitating forces do not travel through emptiness; they come from a localized product of the varying density of the internal and surrounding magnetic monopoles in free space. In the words of the Chinese sage Chang Tsai:

> When one knows the great void is full of ch'i, one realizes that there is no such thing as nothingness.[19]

When one knows that the great void is full of magnetic particles one realizes that there is no such thing as emptiness. Emptiness is subjective. In the words of Ashvaghosha:

> Be it clearly understood that space is nothing but a mode of particularization and that it has no real existence of its

own... space exists only in relation to our particularizing consciousness.[20]

Just because space is subjective or a "particularizing consciousness" does not mean that you can distort it objectively. What is left is the distortion of substance and not space. If relativity would abandon the idea that space has objective reality, this would be acceptable even to the Taoist. Once this is understood, absolute time can be derived from this subjective space in terms of defined quantities, but from matter, time is relative in that when the ruler expands under heat the measurement also changes.

Mach and Einstein's terms seemed to generate models that described a duality and a paradox. By turning things upside down, relativity first moved subjective space into the objective and then quantum probability reversed the direction half way by making the observer the cause of objective reality. The quantum relativist totally denies reality by claiming it incomprehensible. This same condition resides in the medieval and misinterpreted Hindu concept of Spirit. Both the medieval Spirit and the Hindu Spirit 'Brahman' are said to be incomprehensible and incapable of being reasoned about. They are unthinkable. Similarly, when God becomes incomprehensible, God becomes unthinkable. This unthinkable incomprehensibility arrives on the scene more by poor definition of terms than by a lack of understanding.

> The word Brahman is derived from the Sanskrit root brih – to grow – and thus suggests a reality, which is dynamic and alive. In the words of S. Radha. Krishnan, "The word Brahman means growth and is suggestive of life, motion and progress." ...The order of nature was conceived by Vedic seers, not as a static divine law, but as a dynamic principle which is inherent in the universe. This idea is not unlike the Chinese conception of the Tao "the way" as the way in which the universe works; i.e., the order of nature.[21]

The Spirit of God, simply put, is "Godly action." Those who move with universal law and not against it, move with God. Confusion comes when the Spirit of God becomes the objective God. The Spirit of God is right action. It can reside in any person as a degree of right movement. The objectivity of God must yield position separate from all that is not God. The same relationship resides between matter and energy. Energy and matter exist as two separate realities. The ultimate in subjective reality is intelligence. It comprehends the whole picture by separating the predicate reality of observation from objective existence.

The solid material objects of classical physics have been dissolved into relationships of waves of electric and magnetic fields. Probability has become a probability of relative action of existence that comes and goes, but existence itself does not appear and disappear. It changes. Contrary to this essence of classical and Eastern philosophy, modern relativity requires that existence come and go. In the macro sense, galaxies do come and go, but in the atomic sense, particles come and go only in terms of movement from one place to another. It is not a disappearance and reappearance other than light leaving one electron and appearing with another. In-between it was still a piece of objective reality.

But the *summon bonum* (supreme good) of this whole thesis is: galaxy, electron and photon form variable composites of the same material relationship between two physical realities that never appear and disappear. The action between the fields creates the observed reality. Probability defines the change in field configuration and not existence of those fields. The problem with the modern physicist builds when he or she concludes that without motion there is no existence. For this reason the magnetic field appears as some sort of suchness instead of a finer objective reality than that of the electric field.

In conclusion, either the electric and magnetic field differ as manifestations of the same substance or they act closely in a physical way to a third undetermined and un-measurable substance. Perhaps the photo-electric or magnetic monopole particles do not describe the smallest units of objective reality. To say so would encourage the same premature defining as when the atom was named. Objectivity must be carried to infinity. Not that unity will ever be reached, but that the end must be objective and that probability only describes the statistical action between all objective units whether known or unknown.

When all is said and this thesis comes to an end or when a new experiment with the Special Relativity Principle is completed, it might well be illustrated that objects do shrink objectively in the direction of motion. But reason and existence must be maintained by understanding that the observer does not squeeze the object. The relative relationship between the magnetic field density and the electric fields do the squeezing. The faster the object travels, the more magnetic particles the electric fields encounter, thus squeezing like a contracted spring each and every electric field that travels with its axis in the direction of motion. It should be understood that Einstein simply took Maxwell's field equations to determine the squeezing of objects in motion. Einstein's epistemology confused the issue, not the equations. He assumed the constant velocity of light to all observers as perhaps did Maxwell. He forgot the field geometry of Faraday which Maxwell's equations compliment.

Instead of looking for the simple, modern science changed the meaning of space and time to fit observed fact. Conservation was laid bare and curve fitting to the new definitions became the only model for reality. Pre-classical and true religious philosophy was lost in the shuffle. God also disappeared. When the redemption of man comes it will be as if the mote was removed from his eyes. When the time of the end raises, it well be like a new renaissance. When knowledge covers the earth in proper form rather than heaving itself beyond its bounds we are apt to learn that which was from the beginning rather than concoct strangeness to cover our denial. When we realize that the last hundred and fifty years was a long night to science, we can better understand the dawn of a new day.

Conceptual analysis is far more a subjective process than an objective one. It follows the same laws of conservation in reason as inert matter follows the laws of conservation in existence. Intelligence is the subjective application of conservation and matter and light are the objective manifestation of the laws of conservation. Proper conservation of predicate meaning will maintain this agreement.

Samuel Dael

14. Theory Update

> When one seeks the truth and eventually understands, the world will often dismiss it in favor of established norms. Changing conventional knowledge in religion or science is like moving mountains. Tradition is stronger than the cords of death and the reason that knowledge resists discovery for so many years. Politically established authority will always set the way, the egos will control and the common folk will revere every truth turned upside down. The wise are apt to ask, "Where are the seers."
>
> Samuel Dael

Personal Note

I remember reading about the story of an engineer that was hired to design a fabric mill machine. He was told what was wanted, but he could not look at any other current machines available. He had no experience in the design of weaving machines, but nonetheless he was a very good engineer. He ended up designing a machine that revolutionized the industry. I do not know if this is true, but the story is nonetheless interesting.

As a technical writer, my own experience with engineers has been very interesting. I find them generally egocentric, but when left to their vision they will often develop revolutionary solutions. When controlled they maintain the status quo and give exactly what is asked of them. Far too many corporations seek to control the market using engineers like one would use a football. The vision is all in the market and not in the value of the product. Because of this we do not get what might improve life but rather we get what enslaves us. Value is traded for profit that eventually bring self-destruction to the corporation because they cannot see far enough into the future. They only react to the past and present trends. Survival eventually falls into the hands of a leverage buyout in order to control competition, reduce costs and eventually end revolutionary design. In a similar way I think we would be surprised at the number of inventions that are deceitfully purchased only to hide a revolution that could aid the future. Corporate giants endeavor to maintain control over their market rather than seek to fulfill human need. Eventually they all lose.

It does not matter whether one is a great engineer, a physicist, a revolutionary educator, a political constitutionalist, a great prophet, priest or king, the power of those in control determine the outcome rather than the bright minds. Those in control determine what should be accepted and what should be passed to the next generation. Those in control simply satisfy popular dreams rather than seek for human benefit. New ideas are rarely accepted unless they come from those in power and given out for gain. When a fresh approach does come fourth it is attacked through intimidation and when it becomes popular someone will come and overtake it with flatteries by turning the vision into false promises and satiating novelty—eventually turning what was revolutionary into an upside down novelty for the future. The masses are their own demise and those in control feed off their ignorance. When the blind lead the blind both will eventually fall into the ditch of darkness. The many will look for a magical way out, but all that comes is desolations and darkness because the people cannot accept true vision.

Plato changed what Socrates revealed, Paul used new terms to express the philosophy of Christ which in turn were turned upside down by later Christian authorities, and Brigham Young turned the revelations of Joseph Smith into various false doctrines of carnal reward. These examples I mention because the originators where masters of revolutionary ideas. They were martyrs for their revelations, and we have their works only through traditional change and the ever encroachment of darkness. I am sure there are many more who tried desperately to help us see out of darkness but tradition eventually overtakes any original intent. The fathers of the original United States Constitution were a gathering of great minds, but look what tradition has done to the original intent of state responsibility. Politicians changed the constitution through flatteries. We the people could not see what was coming.

Now Einstein was not a martyr, but his contemporaries ran with relativity with such vigor that I call their view the Einstein Illusion. Einstein tried to stem the tide, but could not do so. I cannot help but feel that his work on a unified field theory would have corrected every illusion, but he was now part of the tradition and part of the control. Einstein has seen how things change from original intent and was not about to challenge the evolution of relativity. Such is the case with most of us. We know that to challenge tradition is death, but we also know in our hearts that tradition is somehow more in love of darkness than light.

I do not feel that Socrates, Christ, Joseph Smith, Einstein or the Founding Fathers of the U.S. Constitution would accept what the following generations did to their works. Einstein saw the change before his eyes and could not correct it. When he tried it was in desperation and it become necessary to withdraw his correction. In this Einstein was neither a master nor a prophet. Joseph Smith predicted desolations upon his own house for rejecting a fundamental law. Christ did not live to see his works turned upside down but he did predict what man would do what he historically did.

With these thoughts in mind I would like to review some current theories manifesting more traditional darkness.

Antimatter

Antimatter has been around for some time, but in September 2006 a headline read, "Antimatter discovery could alter physics." The article said that immediately some 13 billion years ago equal amounts of matter and antimatter formed. Much of it quickly acted to annihilate the other, but for little understood reasons a bit more matter than antimatter survived, providing the universe with the planets, stars and galaxies visible today. This artificial balancing act is not much different than the black hole philosophy of coming up with white holes on the inside creating a dual universe. These speculations alarm me because one will back up 13 billion years and say something about reality, but they cannot study anything about what intelligence is and if it is separate from electric matter. They cannot resolve particle and field duality or even the geometrical structure of light—let alone find a better understanding of how the magnetic field and matter affect each other. These things would prove to be far more valuable than a thirteen billion year old theory. For all that scientists have learned about the universe the desire for the mysterious shadows the desire for understanding of what lies before us and what lies ahead. History is essential to learn how it repeats itself, but when past history is not the same model as the future, of what value is it? It does not say whether we will implode or continually explode. There is more conjecture in this than most anything ever written. What has been written in this thesis is certainly justified.

    Antimatter talks of the annihilation of particles where antimatter travels between the real world of matter and the spooky realm of antimatter three trillion times a second. Photons have been annihilating matter through radiation since the beginning, yet we do not call it antimatter. Antimatter is a ghastly term. What if anti electrons are simply momentary polarized electrons that will become annihilated if they come in contact with a regular electron spinning in the wrong direction relative to the surrounding magnetic flux? The result would be various photons spiraling off. Both the electron and anti electron simply blow apart because of contrary spins, but the components still exist ready to be absorbed into other electrons thus making them more massive. It all depends upon the magnetic density and the surrounding magnetic flux in space. Anti electron sounds stranger than a positron but does not need to be. They could be considered the same.

If particle physics would do a better job of developing analog models rather than using terms that do not fit understood relationships, their work may not seem as important. After a model for an atom was first postulated there has never been a tendency to pursue models further. It seems that we have become like Maxwell. We do not want to put physical reality to the test. We would rather keep things mysterious in order to give value to our theories and keep control of our personal ideas and reality formulas.

Consider Fermilab, located near Batavia Ill. It has always sought experiments that are big and expensive and require huge work forces. Fermilab is a 4-mile circular Tevatron particle accelerator that seeks to find more illusive particles. Fermilab began in 1967 and discovered two of the most fundamental particles, the bottom quark in 1977 and in 1995 the top quark, one of the constituent particles of protons. When the lab seeks after smaller particles and tries to understand the various bits of the atom and perhaps how light is structured, there are great discoveries. But when big labs lust after the creation of massive or elusive particles having very short lives by simply piling up photons upon smaller particles says nothing about what we need to know about a more stable universe. We still know very little about the building blocks of reality that absorb light and radiate it back into space the smaller constitutes of matter. One can build anything strange and see that it will not last. The universe remains undiscovered because we seek for strange gods rather than responsibility to understand what we really need to know. The study of black holes, big bangs and particles that do not exist without manipulation illustrates our search for the mysterious and illusive. Most religions do the same. We like mystery beyond our needs more than understanding of what humanity needs.

Physicists talk of Higgs particles, which imbue matter with mass. If they mean gravitational mass, why such an analogy when we have not found the so-called graviton? If they mean inertial mass, do we not know that light itself imbues an electron with mass? How can they talk of imbuing mass when mass is not clearly understood? When light puts mass upon the electron the theorists still talk of explaining how matter and energy are convertible within a visible universe. Is it not light that builds matter? Energy is simply the mass of light in motion. They have not solved the first principles, yet they seek after strangeness as if they can revolutionize the truth without understanding more of the basics. Why find an illusive particle when we do not fully understand the construction of the electron? When we understand the simple we shall surly see that the many strange, momentary things are nothing but overbuilt photons that disintegrate as fast as they are made. The mystery is nothing but overstuffed fields ready to fly apart for lack of equilibrium, which is nothing but a lack of magnetic structure against the surrounding magnetic space.

The particle physicists and their expensive equipment are forever learning but will never come to the knowledge of the truth. They are still back in the materialist age of solid-like things that are held together by mysterious powers of energy without substance. They think the field is pure energy and have denied Faraday's concept. They use Maxwell's equations to generate electric and magnetic fields but deny their particle-like natures.

The discovery of bizarre particles comes at a time when the future of Fermilab may be forced to shut down if Congress does not approve construction of a multibillion-dollar, 18-mile long International Linear Collider. Is it the need for funds that causes the discovery of antimatter or is it simple things that get complex names to justify the money? When money is spent on new projects to study something worthwhile, the day will always come when the practitioners need to come up with reasons to spend more. It is the nature of every bureaucracy, and science is not exempt. Make your work complex in order to justify your job.

Gravity and Dark Matter

I once told a geologist that travels the world searching for oil about the earth's variable magnetic density at different locations on the earth and thus gravity would also be variable. (He was one who broke some barriers in modern geology for applying a three dimensional methodology in determining and predicting where earthquakes might arise.) His answer to me was that it is entirely possible because the strength of gravity is not the same at every point on the earth. According to Discovery magazine August 2006:

> Your weight is not the same everywhere. Because Earth is not a perfect sphere, the pull of gravity is stronger in some places than in others. It's also in a constant state of change, moving with Earth's mantle, falling sea levels, and even tropical storms. The Gravity Recovery and Climate Experiment mission, better known as GRACE, was launched in 2002 by NASA and the German Aerospace Center to measure exactly how what goes up must come down.

I suggest that the Earth's core and the shifting magnetic poles have just as much to do with the variable gravity intensity or as this thesis says a variable magnetic density. Authority, however, thinks too much in terms of mass rather than the nature of the magnetic field generated by the mass. The gravity map to date has little to do with land or even the more massive equator though authorities give mass as the cause of gravity. The largest weak area is south of the Indian Ocean closer to the equator and the next weakest is directly north over Russia. Massive land has nothing to do with gravity. The strongest is in the Pacific Ocean with less land mass. According to current gravity mapping, what goes up must come down faster in London than in Athens. With the above in mind, land mass does not suggest a stronger form of gravity nor does the equator. Land mass may actually weaken gravity because mountains temporarily divert the magnetic field density away from its normal orientation to the earth's surface. With the ocean and flat terrain, the magnetic field is less contorted and smoother. The variable density of the magnetic field would act more perpendicular to flat surfaces. At the equator a specific magnetic line would not fallow the contour of the earth, but would pass below the ocean surface just north of the equator and exit on the surface further south (assuming a southerly direction—no one really knows). An object a mile above the equator would then be lighter than a mile above the United States. This is relative to the magnetic density and not the Earth's surface. Gravity would be weaker at the equator as gravity maps do suggest and not stronger at the equator because of the bulging mass. One would also have to consider the location of the magnetic poles and the state of the earth's core, but the variable density of the magnetic field is more in line to discovery regarding gravity than following the idea that large masses attract matter at a distance. I think comparing areas near the north and south magnetic poles respectively could give us more. If the theory of gravity expressed in this book is true, then one should see differences that can only be explained by the idea that gravity is a result of the variable magnetic density surrounding the earth and not relative to the center of mass.

I think if we studied the magnetic field and what facilitates it and limits it, understanding may give rise to a better understanding of gravity. I will bet there are many questions about the data currently obtained about gravity but few will admit it. It is the same as when tests were made with atomic clocks moving in a west and easterly direction. The differences could not be explained as to relativistic velocity. No one considered the magnetic fields of the earth and sun, as the clocks moved into the field in the easterly direction and more at rest relative to the magnetic field in a westerly direction.

The great paradox is that gravity was one of the first forces man learned, but still the most elusive. The problem became even more difficult with a term twisting of Einstein's General Theory in 1915. Science now attributes a change in gravity as a change in space. If Einstein could only have determined it to be a change in magnetic space it would not have been so difficult to have eventually found the truth.

LIGO, the Laser Interferometer Gravitational Wave Observatory, is a pair of three-mile-long gravitational-wave detectors in Washington and Louisiana that cost $365 million and took 11 years to build, and yet they may just barely be able to pick up signals from the ultraviolet collisions that give birth to massive black holes. It is amazing what the scientist will do to justify his job. If the scientist truly began to question what causes light to bend in a gravitational field twice that of matter rather than talk of curved space, we might get somewhere with gravity. I do not think it will happen because we have to first determine the nature of light and that is still a contradiction between wave and particle mechanics.

The concept of gravity introduced by Newton has dominated physics. Objects under the influence of the Earth's gravity will fall towards (and thus orbit) the center of mass of the Earth. This is only a mathematical relationship. Irregular surfaces may alter this assumption in very minor degrees. The magical action at a distance is traditionally believed with such strong support that physicists have created dark mater to justify the behavior of spiral galaxies. In many of these galaxies all of the stars in the middle and outer parts of these galaxies orbit with the same speed, in seeming defiance of Newton's laws, therefore physicists postulate dark matter at some point that would cause this strangeness. The outer stars should move more slowly or leave the galaxy due to their high velocity. The spiral arms do indicate a strange condition, but they also indicate a pattern due to some underlying cause that may be due to dark matter or some other principle. Consider that the outer tails are drawn to the density of the arm more than to the center of the galaxy yielding the net velocity almost equal to the stars next to the center. If the outer tails move as fast as stars near the center, it is because they are caught in the magnetic field density of the arms and not the center of the galaxy. It is like falling towards the mass of the arm than towards the center of the galaxy. This illustrates that action at a distance using mass points of the total mass calculation is not appropriate. Matter reacts to the variable magnetic field density surrounding the tails and not to the center of mass of the galaxy. Since the variable density towards the base of the arms increases greater than towards the center of the galaxy, the outlying tails appear to move toward these arms at a similar velocity as those stars on the surface.

There is too much concern about determining the total mass of a galaxy and placing it at the center for calculation. In a spiral galaxy there is a lot of mass far distant from the center with a lot of space between the dense center and the spiral arms. If we get rid of the idea that the center of mass acts upon a far distant star and admit that the magnetic configuration surrounding each star determines its behavior, we might understand that dark matter is a foolish concept. Would not intelligent matter be far more interesting than black holes and dark matter? I guess not because that would imply a greater interest in the intelligence of God.

Newton's gravity alone should be able to explain that matter is denser in the arms than between those arms and the center. The attraction is not only to the center of the galaxy but also toward the arm leading the way. The spiral pattern could be due to the galaxy's magnetic structure and other concepts that can alter the magnetic field density within the arms. Everywhere in nature spiral forms are may be due to the magnetic field density and rotational motion.

From Science Daily (March 26, 2006), a headline read: "Anti-Gravity Effect? Gravitational Equivalent Of A Magnetic Field Measured In Lab." Scientists funded by the European Space Agency had measured the gravitational equivalent of a magnetic field for the first time in a laboratory. Under certain special conditions the effect is much larger than expected from general relativity and could help physicists to make a significant step towards the long-sought-after quantum theory of gravity. The experiment is not clear, but hints that someone is trying. The mention of the undiscovered graviton in the article did not help the situation. If they only considered that the experiment was generating an increase in magnetic density surrounding the very fast rotating ring they might understand, but the tendency is to think in terms of quantum physics and relativity. Few realize that if gravity is quantized it must of necessity reach an end for every large body because the last quantum cannot be divided.

What is needed is more research in local reactions with magnetic fields and the curvature of light. Since light bends around corners it should be understood that it is not attraction but curvature into more dense magnetic properties. The same explains even light refraction. A high speed magnetic ring might simulate a particle and give us more insight into atomic matter, but I do not think that things billions of light years away in both distance and time would do as much for our understanding as if we knew more about light and magnetism and their relationship. We need to get rid of the idea that mass converts to energy and vise versa. Rather mass is converted to light—the photo-electric makeup of mass. Just as we convert angular momentum to linear momentum, the very same happens to light. The difference is not in the photo-electric field, but in the nature of the magnetic field. Light has no circulating magnetic field. A better description would be that light spirals around a magnetic field at rest. This does not change the relationship or the nature of the photo-electric field as light or in matter. It is the same. Only the state of magnetic field differs. The relativistic mind just does not find this as interesting. It probably will not be a means of getting grants or building massive accelerators. It is not strange enough.

Mordehai Milgrom of Israel did not believe in dark matter and has boldly updated Newtonian motion and in doing so, has reconceived Einsteinian gravity. He changed Newton's equations in such away that gravity diminishes differently at long distances. This would be entirely agreeable with the idea that the earth's magnetic field diminishes differently. It would also explain that there was no need to create dark matter in order to explain things at very large distances in the same way we consider things close. We should understand that light travels faster in a vacuum in far distant reaches of outer space than in a vacuum on the surface of the earth or sun. So too would gravity react disproportionably? The idea that gravity is infinitely graduated just does not make sense. When it comes to the last line of magnetic density of a particular planet or particle, beyond that point is zero gravity. If the magnetic field is quantized and also the cause of gravity then a planet light years away would have absolutely no effect on earth. Gravity is not infinitely large according Newton. It is zero at certain large distances. Gravity determines local motion as long as there is a variable magnetic density. The Universe cannot attract and fall into itself no matter how slow the so-called expansion becomes. Once the corrections of gravity are made we will find that the universe is dynamic, but not expanding.

String Theory

String theory is a work in progress, so trying to pin down exactly what the science is, or what its fundamental elements are, can be kind of tricky. The key string theory features include:

- All objects in our universe are composed of vibrating filaments (strings) and membranes (branes) of energy.
- String theory attempts to reconcile general relativity (gravity) with quantum physics.
- A new connection (called supersymmetry) exists between two fundamentally different types of particles, bosons and fermions.
- Several extra (usually unobservable) dimensions to the universe must exist.

There are also other possible string theory features, depending on what theories prove to have merit in the future. Possibilities include:

- A landscape of string theory solutions, allowing for possible parallel universes.
- The holographic principle, which states how information in a space can relate to information on the surface of that space.
- The anthropic principle, which states that scientists can use the fact that humanity exists as an explanation for certain physical properties of our universe.

- Our universe could be "stuck" on a brane, allowing for new interpretations of string theory.
- Other principles or features, waiting to be discovered.

I find it appalling for modern physics to search for singularity in the popular string theory. It is true that physics has found 57 varieties of particles and it is natural to seek for a common underlying element, but first a distinction has to be obtained. That distinction is the common relationship of all particles and that is described in the photo-electric and magnetic field relationship. The more complex the relationship and the more dense the fields, the larger the particle the physicist will find. This is based upon the unified field theory and has been neglected since the death of Einstein. To search directly for singularity defies distinction and opposition in all things. Opposition does not appear out of the same substance without first finding a distinction between at least two particles.

String theory imagines that every substance in the universe is made of just one thing. The theory postulates that tiny vibrating strings at different resonances create 57 different particles and everything else. String theory requires at least 9 spatial dimensions, six of which are not noticeable to those of us living in a three-dimentional world. At the moment the theory has no experimental support. No one has seen strings. They would be way too small, by a factor of many trillions. As for the hidden dimensions, they would be wherever you want them.

I would not want to poke fun at the nine dimensions in string theory because a dimension to a mathematician can be somewhat different than to a geometrician. In traditional ways the mathematical description of space in physics is a point particle description as if all the mass was at a point. By extending this to a consideration of an infinite number of points grouped together in a particular manner intensifies the difficulty. In string theory the various dimensions are described as the Kaluza-Klein idea of curled up dimensions having different sizes, shapes and numbers of holes. String theory allows two mathematically distinct curled up spaces to yield both a traditional physical model as well as a string model. So too does the electric and magnetic relationship. The electric and magnetic relationship is two curled fields with a traditional center point to place all of the mass. Why do string theorists need new terms where classical terms would do fine?

In string theory, the model is based on tiny loops and hence differs markedly from the mathematical mass point description. This, in turn, allows two mathematically distinct curled up spaces (I would call fields) to yield physically identical models to traditional mass point physics. String Theory is a purely theoretic phenomenon which relies profoundly on the extended nature of a string or can I say the extended nature of a field. The extended nature is where mathematical physics gets the nine dimensions. To better understand the particle you cannot simply place a point, you must be able to mathematically define the complex field relationship. String theory does this with mathematical derivatives and not with the mind's eye dimensional view.

The real problem with String theory is that it reduces all things to one reality as if both the magnetic and electric fields are of the same string-like origin. String Theory is a philosophy of singularity—not unlike modern either theories. Ether theories were debunked along time ago because the Ether was considered to be a rigid substance. I am sure that seeing things as curving, looping and curling required that one could not use the old ether concept. The physicist now looks for a more dynamic process in the word string. What was wrong with the word field? Field theory does not provide for singularity of one single thing that permeates all of space—something modern thinking wants so desperately to prove. It is psychologically the same as proving that there is no good and evil. It is a safe way of thinking because opposition is created out of nothing—as when you create an electric current you simultaneously create a magnetic current out of empty space. Faraday's field theory said that space is not empty. Empty thinking started with Maxwell and still exists in modern physics. Even particles are energy loops or as the ether theorists would say, "Ripples in the ether." Ether theories are a singularity of something. String theory is a singularity of nothing other than pure energy. Pure energy comes from $E=mc^2$ as if energy is converted to mass and back again. We forget that the equation says that if *mass=0* then *energy=0*. Energy needs mass to be energy. It cannot exist of its own. Energy comes in quantum units because mass comes in quantum units. You cannot destroy mass as you cannot destroy *light*. You can only change its direction relative to the state of the magnetic field.

Physics constantly taunts that there is no substance in string theory as there is no substance in fields also, thus different fields must be composed of one reality called strings—a single component of reality. This need for singularity in strings and also in modern aether theory is a basic psychological problem. It is more desire to create the same illusions as relativity in that all things are figments of chance happenings. Anything one says goes and nothing is wrong and nothing is right. All are justified and the one basking in singularity becomes a God of his own make. At all costs there is a tendency to destroy any difference between two realities, as there is no difference between a subject and the object or more particularly between the object and the meaning of the object.

There must be a dichotomy or differing substances in the universe. You cannot have equilibrium without opposition. You cannot have that which is good without that which is evil. Differences are essential. In fact, it is far more important to establish a third substance than try to make two into one. If we can find a particle of intelligence, a particle of action and better understand the base particle of light we could then learn how to construct anything out of design rather than happenstance. This implies three universes superimposed. They are subjective particles of intelligence, predicate magnetic monopoles, and objective photo-electric matter. Just as electric matter interacts with magnetic space in order to create what we know as physical reality, intelligent matter, following the left-hand rule, just might interact with magnetic space in order to create individual intelligence. Call me mad, but the obvious need for a three-part reality lies in knowing more about the magnetic field rather than breaking it down into strings of energy.

Only with three can you create the proportional need of responsibility and choice. Choice is not a mechanistic happening of a single reality—it is the bringing of two into equilibrium. He who looks for more or less than this has a psychological problem possessing an intelligence having an exalted ego. He denies the nature of right action, the nature of God and the nature of responsibility—all of which are derived from the essence of intelligence or the third reality.

References

Introduction

1. Richard P. Feynman, "Classic Feynman," page 19.
2. Richard P. Feynman, "Classic Feynman," page 19.

1. The New Reality

1. Ernest Becker, "The Denial of Death," p. 47
2. Ernest Becker, "The Denial of Death," p. 47
3. Ronald W. Clark, "Einstein The Life and Times," p. 385
4. Michael Talbot, "Mysticism and the New Physics," p. 161
5. Ernest Becker, "The Denial of Death," p. 280
6. Ernest Becker, "The Denial of Death," p. 27
7. Ernest Becker, "The Denial of Death," p. 17

2. Striking a Balance

1. Bertrand Russell, "The ABC of Relativity," p. 138
2. Ernest Becker, "The Denial of Death," p. 171
3. Ernest Becker, "The Denial of Death," p. 170
4. Carl Sagan, "Cosmos," p. 262

3. The Square of Reason

1. Isaac Asimov, "Understanding Physics: The Electron, Proton, and Neutron," p. 13
2. Pyramid Its Devine Message by D. Davidson and H. Aldersmith Vol. 1, London, Williams and Norgate, LTD. 1927
3. Arthur M. Young, "The Geometry of Meaning," p. 16

4. The Geometry of Mathematics

1. Isaac Asimov, "Understanding Physics: Motion, Sound, and Heat," p. 17
2. Philip Davis & Reuben Hersch, "The Mathematical Experience," p. 341

5. Much About Nothing

1. Arthur M. Young, "The Geometry of Meaning," p. 147
2. Hermann Bondi, "Relativity and Common Sense," P. 58
3. Albert Einstein, "The Meaning of Relativity," p. 30
4. Albert Einstein, "The Meaning of Relativity," p. 31

6. The Einstein Illusion

1 Hermann Bondi, "Relativity and Common Sense," p. 120
2 Martin Gardner, "The Relativity Explosion," p. 129
3 It is now over 100 years. The comment was first written in the early 1980's
4 Ibid
5 http://en.wikipedia.org/wiki/Photoelectric_effect

7. Objectivity in Matter

1 Albert Einstein & Leopold Infeld, "The Evolution of Physics" p. 31
2 Isaac Asimov, "Understanding Physics: The Electron, Proton and Neutron," p. 19
3 Isaac Asimov, "Understanding Physics: The Electron, Proton and Neutron," p. 20
4 Ibid., p. 58

8. Predicate Nature of Energy

1 Isaac Asimov, "Understanding Physics: Motion, Sound and Heat," p. 84
2 Ibid., p. 38
3 Ibid., p. 30
4 Ibid., p. 33
5 Nigel Calder, "Einstein's Universe," p. 102
6 Robert Geroch, "General Relativity," p. 166
7 Nigel Calder, "Einstein's Universe," p. 105
8 Ibid., p. 94
9 This manuscript was first completed in the fall of 1987, but began eight years earlier.

9. Relativity and Epistemology

1 Ayn Rand, "Introduction to Epistemology," p. 99
2 Nigel Calder, "Einstein's Universe," p. 55
3 Nigel Calder, "Einstein's Universe," p. 56
4 Nigel Calder, "Einstein's Universe," p. 58
5 Ibid., p.61
6 Arthur M. Young, "The Geometry of Meaning," p. 16
7 Nigel Calder, "Einstein's Universe," p. 33
8 Richard P. Feynman, "Classic Feynman," page 19.
9 Nigel Calder, "Einstein's Universe," p. 35
10 Weidner/Sells, "Elementary Modern Physics," p. 346

10. Action at a Distance

1   Heinz R. Pagels, "The Cosmic Code", p. 232.
This quote is almost thirty years old and does reference the period of the sixties and seventies. Field Theory was more popular then than now.

2   Nigel Calder, "The Key to the Universe", p.63
3   Heinz R. Pagels, "The Cosmic Code", p. 232
4   Timothy Ferris, "The Red Limit", p. 42
5   Heins R. Pagels, "The Cosmic Code", p. 237
6   "The Penguin Dictionary of Physics", p. 230
7   http://en.wikipedia.org/wiki/TeV#Photon_properties
8   From the Internet-no source given
9   Isaac Asimov, "Understanding Physics: Motion, Sound, and Heat, p. 226
10 Isaac Asimov, "Understanding Physics: Motion, Sound, and Heat", p. 225
11 Earnest Becker, "The Denial of Death"
12 St. John 3:8
13 S.E.Shnoll, K.I. Zenchenko, I.I. Berulis, N.V. Udaltsova, I.A. Rubinstein (2004). "Fine structure of histograms of alpha-activity measurements depends on direction of alpha particles flow and the Earth rotation: experiments with collimators". arXiv:physics/0412007 [physics.space-ph].

14 Stanford University (August 2010). "The strange case of solar flares and radioactive elements". Press release. http://news.stanford.edu/news/2010/august/sun-082310.html

15 C. Giunti, C.W. Kim (2007). Fundamentals of neutrino physics and astrophysics. Oxford University Press. p. 255. ISBN 0198508719. http://books.google.com/?id=SdAcSwTR0CgC&lpg=PA255&dq=majorana%20neutrino%20helicity&pg=PA255

16 J.N. Bahcall (1989). Neutrino Astrophysics. Cambridge University Press. ISBN 052137975X.

17 Davis, D. Raymond Jr. (2003). "Nobel Lecture: A half-century with solar neutrinos". Reviews of Modern Physics 75 (3): 10. doi:10.1103/RevModPhys.75.985

11. The Big Illusion

1   Timothy Ferris, "The Red Limit", p. 151
2   Carl Sagan, "Cosmos", p. 237
3   Isaac Asimov, "Understanding Physics: Light, Magnetism and Electricity." P. 154
4   "Particle creation by black holes". Project Euclid. http://projecteuclid.org/Dienst/UI/1.0/Summarize/euclid.cmp/1103899181. Retrieved 19 May 2008.

12. Maxwellian Space

(Endnotes)

1. Nikola Tesla Electrical Review and Western Electrician, July 6, 1912
2. "Electromagnetism, Maxwell's Equations, and Microwaves". IEEE Virtual Museum. 2008. http://www.ieee-virtual-museum.org/exhibit/exhibit.php?id=159265&lid=1&seq=3. Retrieved on 2008-06-02.
3. J J O'Connor and E F Robertson, James Clerk Maxwell, School of Mathematics and Statistics, University of St Andrews, Scotland, November 1997
4. James Clerk Maxwell, A Dynamical Theory of the Electromagnetic Field, Philosophical Transactions of the Royal Society of London 155, 459-512 (1865).
5. Nahin, P.J., Spectrum, IEEE, Volume 29, Issue 3, March 1992 Page(s):45
6. http://www.arrl.org/tis/info/catv-ch.html as of 1/3/09
7. New York Herald Tribune, September 11, 1932
8. http://peswiki.com/index.php/PowerPedia:Tesla's_Dynamic_Theory_of_Gravity
9. http://philsci-archive.pitt.edu/archive/00002875/01/Torretti%E2%80%A2Crit._&_Adv.pdf

13. A Matter of Intelligence

1. John Taylor Gatto, "Underground History of American Education", p. xxiii.
2. Timothy Ferris, "The Red Limit", p. 151
3. Albert Einstein, "Ideas and Opinions", p. 48
4. Ronald W. Clark, "Einstein The Life and Times", p. 38
5. Ronald W. Clark, "Einstein The Life and Times", p. 38
6. Ronald W. Clark, "Einstein The Life and Times", p. 37
7. Robert H. March, "Physics For Poets", p. 107
8. Ronald W. Clark, "Einstein The Life and Times", p. 38
9. Robert Jastrow, "God and the Astronomers", p. 17
10. Robert Jastrow, "God And The Astronomers", p. 17
11. Philip J. Davis and Reuben Hersh, "The Mathematical Experience", p. 368
12. Fritjof Capra, "The Tao of Physics," p. 21
13. Fritjof Capra, "The Tao of Physics," p. 29
14. Fritjof Capra, "The Tao of Physics," p. 42
15. Fritjof Capra, "The Tao of Physics," p. 93
16. Fritjof Capra, "The Tao of Physics," p. 103
17. Fritjof Capra, "The Tao of Physics," p. 105
18. Fritjof Capra, "The Tao of Physics," p. 195
19. Fritjof Capra, "The Tao of Physics," p. 209
20. Fritjof Capra, "The Tao of Physics," p. 150
21. Fritjof Capra, "The Tao of Physics," p. 176

Index

A

absorption lines 290-1, 309

acceleration 55-8, 60-1, 71, 126, 129, 141, 146, 168-9, 173, 180, 186, 223, 251

aether 99, 101-2, 119, 157, 248, 253, 262, 264

aether wind 99-100, 102, 104, 254

alchemist 32, 133

analog, intuitive 321, 324

analogical 314, 316, 324

analogical thinking 313-15, 317

analysis of observation 314, 317

Analytical thinking 313-15

angular ix, 39, 138, 141, 150, 170-2, 174-5, 236

angular energy vii, 62-3, 138, 165, 172, 175-8

angular inertia 168, 170

angular momentum

  conversion of 138, 140, 149

  photo-electric field's 193

angular velocity 165, 168, 179-81, 192, 202-3, 222

antenna 248-9, 252, 254-61, 282

antimatter vii, 347-8, 350

apparent density 131, 164-5

Asimov, Isaac 41, 65, 145, 361-3

atheism i, 96, 99, 319-20

atomic clocks 111, 155, 162-3, 165, 236, 262, 289, 352

atomic magnetic field structure 221

atomic weight 133-6, 204-5, 210

Avogadro's number 135

axioms 6-8, 12-15, 17, 28-9, 32, 35, 37, 45, 65, 133, 165-8, 298

  balanced 13-15

  self-evident xix, 6, 8, 17, 34

axis 7-8, 55, 58, 110-11, 145, 167, 169, 175, 191, 193-4, 202, 205-6, 209, 215, 258, 262

B

background radiation 286, 292, 294-5, 308, 311, 326

battery 198, 202, 216, 266-7

Becker, Ernest 361

Big Bang 277-9, 291, 294, 296, 311, 319, 325-8, 331

Big Illusion vii, 277-8, 280, 284, 286, 288, 290, 292, 294, 296, 298, 300, 302, 304, 306, 308

binary stars 289, 291-2

black body 293-5, 310-11

black holes ix, 23, 32, 63, 293, 310-11, 314, 328, 349, 354

bond

  double 218-20

  single 217-18, 220

British inch 53

C

Calder, Nigel 362-3

causality 19, 26, 31-2, 323

changed Newton's equations 355

Changing Magnetic Density 309

charged particles 95, 179-82, 188, 209, 249-50, 265, 268

Christ 319, 346-7

circulating magnetic field 187, 194, 203, 209, 213, 219, 222, 355

Clark, Ronald W. 361, 364

Classic Feynman 361-2

classical relativity 20, 86, 101, 109, 112

clocks 51, 70-1, 80-1, 83, 85, 87, 116, 162-5, 289, 352

color shift 106, 108, 110-11, 118, 279, 288-90, 292, 304-7

color temperature 293-4

conceptual terms 41-5

conductor 245-6, 249-50, 254, 256-61, 263-5, 267-8, 270-1

conservation 7, 13-15, 18-19, 23-4, 26-9, 35, 37-8, 52-3, 59-67, 132-3, 146, 165-7, 171-2, 227-8, 239, 297

  law of 60, 62-4, 67, 132, 134, 165, 231, 239

  laws of 63, 155-6, 179, 183, 233, 235, 278, 318, 320-1, 331

Conservation of gravitational MASS 60

Conservation of inertial MASS 60

Conservation of Mass vi, 132

constant, cosmological 324-6

constant velocity xviii, 38-9, 59, 73-5, 82, 90, 96, 99, 104, 112, 115-16, 119, 146, 153-6, 190, 295

contraction 101, 111, 115-19, 219, 293

conversion

  mass energy 149

  mass-energy ix

Cosmic Code 363

cosmologists viii, 275, 279, 288, 294-5, 318

curve fitting 317-18, 321, 325, 333, 343

curved space gravitation 142

D

Dalton 134-5

dark matter vii, 280, 351, 353-5

degrees Kelvin 293-4

Democritus 132-4

denial i, 8-9, 16-17, 20-1, 26, 30, 32, 64, 68, 156-7, 229, 295, 297-8, 315-16, 319-22, 324

dense magnetic field 113, 116, 156, 308

dense nuclear magnetic fields, very 176

denser magnetic field 163-4, 251, 270, 285

density 39, 96, 98, 103, 109-12, 116, 126-7, 129, 131, 141, 152-6, 164, 194-5, 203, 212-14, 254-6

flux 189, 193, 215, 224

derivatives 58, 72-3, 169-72

determinism 22, 24, 30

determinist 17, 22-3, 327

dimensions 33-4, 36, 50-1, 55-6, 58, 69-70, 72-3, 82, 105, 172, 229, 240, 245, 254, 356-8

direction, negative 81, 83, 100, 114-15

distinction 41-53, 64-6, 69, 165-8, 185, 226-8, 272, 328, 332, 357

Doppler Effect 83, 91, 105-12, 114-17, 224, 281, 287-8, 291

Doppler Shift 91, 107, 288, 291

duality

  particle/wave 288

  wave-particle 96, 299

E

Earth to Mariner spacecraft 153

earth's magnetic field 107, 117, 192, 223, 226

earth's magnetic field diminishes 355

Eastern mysticism 334-5, 337, 339

Eastern Philosophy 323, 333, 337-8, 342

Einstein iv, x-xi, xv-xvi, 31, 59, 86, 99-103, 111-12, 115, 119-21, 149-52, 161-2, 183, 321-8, 335-7, 347

  Albert 18, 120, 361-2, 364

Einstein contraction 112, 114

Einstein Curve 114

Einstein effect plots 114-15
Einstein equation 90, 115
Einstein illusion and Maxwellian space 275
Einstein relativistic equation 151
Einstein's Clock vii, 161-2
Einstein's equation x, 87, 113, 119, 151-2
Einstein's field theory 118
Einstein's God 325, 328
Einstein's observer paradox 102
Einstein's Photoelectric Effect xii, 106
Einstein's photon 96
Einstein's relativity 20, 23-4, 38, 69
Einstein's relativity principle 65
Einstein's transformation equations 91, 103
Einstein's Universe 18, 362
electric density 190, 255-6, 286
electric field spiraling 103, 284
electric field spirals 104, 117, 291
electric matter 103, 112, 202, 219, 227, 232-4, 256, 287, 348, 359
electric particle field contracts 116
electricity 188, 195, 198, 200, 203-4, 210, 213, 216, 240, 242, 252-3, 363
electromagnetic relationship 112, 128, 214, 216, 248, 257, 274, 301
electromagnetism 100, 119, 185-6, 237-8, 244, 253, 364
electron field 209
Electron in Magnetic Space 207
electron levels 216, 289
electron spin 206, 210
electron volts 198-9, 267
electrons 38, 108-9, 128, 137-40, 150, 190-4, 196-222, 225-7, 236-8, 240, 243-5, 265-8, 270-2, 284-5, 350, 361-2
 polarized 210, 348
electrons emit photons 289
element chart 137, 204-5, 210-11, 221
empty expanse ix
empty space 59, 67, 96, 98-9, 108, 116, 119, 140, 151, 158, 185-7, 192, 252-4, 294, 358
energy xv, 60-3, 120-1, 137-9, 145-50, 156-8, 160-1, 169-78, 180-4, 187-90, 196-200, 242-8, 250-3, 255-6, 293-5, 359-60
 conservation of 106, 146, 233
 measure of 198
 total 172, 174-5, 273, 293
 zero 184, 244
epistemology vii, ix-x, xv-xvi, 1, 12, 21, 149, 152, 161-2, 164, 166, 168, 170-2, 174, 176, 182-4
equilibrium 6-7, 9-10, 12-13, 24, 29, 33, 60, 63, 140, 201, 203, 235, 239, 271, 293, 359-60
errors, relativistic timekeeping 162
evil 21, 314-15, 358-9
exchange photons 178, 250
existence xix-1, 4-8, 10 16, 19-19, 23-4, 26-8, 30, 32-5, 37-8, 60-1, 189-91, 232-4, 252-3, 321-3, 339-40, 342-3
 changing 232
expansion, universal 291-2, 296, 325-6
experience i, 15, 29-30, 65, 126, 130, 207, 210, 289, 345
experiments, slit 157-8
exterior magnetic field 267

F

faith vi, xii, xix-1, 6-12, 14-22, 24, 27, 32, 41, 63, 65-6, 128, 130, 232, 313, 320
 blind 8, 11, 20, 124

Faraday xv-xvi, 35-6, 108, 117, 142, 182-3, 190, 209, 248, 253, 273, 275, 343

Faraday's particle fields xvi, 183

fear x, xix, 9, 15-16, 18, 20-3, 28-30, 34, 64-5, 68, 298, 313, 319, 333, 336-7

feminine 228, 230, 317, 329

Fermilab 349-50

Ferris, Timothy 277, 363-4

Feynman, Richard P. 361-2

field density 102, 119, 121-2, 153, 155, 158-9, 165

earth's magnetic 165, 192

net magnetic 163, 289

variable magnetic 98, 122, 340, 353

field theory 62, 131, 142, 151, 153, 183, 189-90, 192, 197, 205-6, 208, 214, 216, 222, 276, 326

geometrical unified electromagnetic 322

unified iv, x, 132, 186, 189, 226, 322, 325-6, 347, 357

fields, electromagnetic 141, 252-3, 284, 364

fireball, primeval 292, 295, 327-8

Fitzgerald 99

Fritjof Capra 334, 364

G

galaxies 94, 109, 121, 278-9, 281, 292, 295, 297, 330, 342, 348, 353-4

General Theory of Relativity xv, 94, 163, 324-5

geometrical center 147-8, 187

geometry 12-13, 46, 55-6, 74, 81, 83, 130-1, 151-2, 156, 188, 191-2, 216, 229, 253-4, 268, 285-6

Geometry of Mathematics vi, 65-6, 68, 70, 72, 74, 76, 78, 361

gluons 186-7, 208

God i, vii-viii, xi, xviii, 16, 23, 232, 234-6, 239, 262-3, 296, 313-15, 317-33, 339, 341-3, 359-60

existence of 312, 330, 332

objective existence of 332

objectivity of 232, 329, 332, 342

God of magic 239, 321-2, 333

God's existence 330-2

gravitation 11, 60-1, 96, 118, 126-7, 130, 132, 186, 223-4, 226, 292

gravitational constant 127, 129-31

gravitational ENERGY 60, 146

gravitational equations 11, 268

gravitational field 39, 71, 94-5, 110, 121-2, 126, 150-1, 165, 246, 251, 353

gravitational mass 60-2, 72, 122, 131, 190, 350

gravitational mass 71

gravity 10-14, 39, 61, 94-5, 126, 128-9, 153, 185-7, 223-4, 242-3, 250, 264-5, 270, 272-3, 311, 351-6

gluon of 186-7

theory of 28, 352

gravity and dark matter vii, 351

gravity maps 351-2

ground orbit 211-12

H

harmonic iv, 47, 228, 230, 329

harmony 7-8, 10, 140, 231, 235, 238-9, 313-14, 329, 338

human honesty 319-20

I

immortality 17, 21-2, 30, 32, 64, 297, 320

indeterminism 22, 24, 29

inequality 42, 313

inertia 60-2, 70, 126, 146, 168-71, 174, 187, 189, 226, 251, 340

  kinetic 168, 170

  total 170-1

inertial 60-2, 131, 138, 146

inertial mass 60-3, 71, 131, 190, 350

  understanding of 130, 132

infinity xiii, 46, 115, 228-30, 246, 248, 254, 274, 308, 311, 314, 316, 319, 331, 343

intelligence vii-viii, xi, 5, 8, 64, 67-8, 130-1, 171, 226-34, 238-9, 313-16, 320-2, 328-30, 332-4, 342, 359-60

  intuitive xi

  supreme 232, 323, 331-2

intelligent understanding viii, 53

Intuitive leaps 316, 318, 322

invariance 2-3, 45, 52

invariant 2-4, 9, 45-6

isotopes 135-6

J

justice 7-8, 16, 19, 23, 231, 314, 329

K

kinetic energy 138, 146-8, 168, 170, 172, 174-7, 180-1, 198, 244, 247

L

Laser Interferometer Gravitational Wave Observatory 353

laws, first 146-7

light photon particle 88-9

light's velocity constant 38

linear energy 62, 138, 172, 174-5, 308

logic 25-6, 29, 33-5, 59-60, 64, 66, 131, 178, 193, 230

Lorentz 100-1, 103

Lorentz equation 113-14

Lorentz transformation 100-1, 113, 115, 183

M

Mach 340-1

Mach's philosophy 340

magic ix, 7, 15, 18, 21-2, 28, 32-3, 119, 128-30, 138, 185, 232, 239, 244, 321-2, 336-7

magnetic axis 164, 190-1, 203, 208, 223-5

magnetic density 94-5, 109-12, 117-18, 122-3, 152, 155, 159-60, 192-3, 201, 218-19, 222-4, 261-2, 266-7, 271-3, 283, 286-92

  apparent 115, 289

  graduated 95, 127, 158, 270, 273, 326

  variable 203, 251, 307, 311, 351-2, 356

magnetic field 103-4, 107-13, 115-20, 138-43, 163-5, 178-9, 181-3, 186-90, 192-7, 200-5, 207-10, 213-16, 223-6, 247-56, 259-62, 350-2

  rotating 98, 245, 271

  static 241

magnetic field density 102-3, 111, 115-16, 119, 121-2, 142, 150-1, 155-8, 160, 162, 165, 196, 240, 246-7, 291-2, 353-4

magnetic flux 191, 193, 201-4,

Index | **370**

207, 215, 221, 226, 266, 270-1

magnetic medium 110, 116, 142, 243, 254-5, 263-4, 275

magnetic monopoles 189-91, 203, 214, 326

magnetic particle field xv, 122, 140, 284

magnetic particles 116-17, 164, 182, 189, 233, 242, 245, 247, 256, 262, 332, 340-1, 343

magnetic space 96, 108-9, 111-12, 116-18, 140-1, 152-3, 155-6, 186-8, 190-2, 206-8, 245, 248-50, 254-5, 261, 263-5, 272-4

magnetic waves 246, 256, 258-9, 261, 286

magnetism 67, 188, 210, 214-15, 234, 252-3, 273, 355, 363

magnitude, proportion of 49-51, 57-8

masculine 228, 230, 317, 329

mass
  apparent disappearance of 63, 140
  atomic 137, 165
  center of 72, 129, 147-8, 169, 174, 176, 187, 189, 191, 352-3
  conservation of 60, 62, 71, 132, 176
  objective 71, 126, 147, 149, 171
  quantum of 149, 292, 295
  zero 138, 184, 236, 243-4

mass control 58, 168

mass/energy 143

Mass-Energy vi, 137

mass/energy reality 63

mass particles 106, 139, 150, 169, 244-5, 294

mass relativity 71, 179

mathematical zero point 77, 86

matter xv-xvi, 11-13, 125-8, 130-4, 138-42, 151-3, 176-8, 232-4, 237-8, 261-3, 286-7, 294-6, 308-11, 340-2, 348-50, 352-6
  intelligent 354, 359
  objective photo-electric 359

matter attracts matters 11, 13, 151

matter of Intelligence vii, 313-14, 316, 318, 320, 322, 324, 326, 328, 330, 332, 334, 336, 338, 340, 342

Maxwell xvi, 35, 115, 117, 119, 182-3, 188, 243, 249, 252-4, 273-5, 326, 343, 349, 358

Maxwellian and Einstein space 249

Maxwellian Space vii, xvi, 96, 241-4, 246, 248-50, 252, 254, 256, 258, 260, 262, 264, 266, 274-6, 325-6

Maxwell's equations xv, 35, 99, 112, 115, 119, 151, 155, 189-90, 248-9, 254, 350, 364

Maxwell's field equations 113, 115, 151, 183, 343

$E=mc^2$ 62, 137, 141, 149, 172, 174, 176, 180, 273, 299, 359

mechanical theory 92-3, 96

medium 42-3, 84, 91-4, 96-9, 105-9, 111-12, 121, 142, 155-7, 177-8, 202, 216-17, 242-5, 247-8, 253, 285-6
  physical 243, 285

metaphysics 26, 335-6

MHz 258-60

Michelson-Morley 99
Michelson-Morley experiment 103, 117, 253, 288
Michelson-Morley experiment 99-102, 154-6
model, helical 285, 289, 292, 301
modern quantum theory xvii, 298, 300, 325
momentum, conservation of 145, 195
monopoles 193, 202, 205, 283-4
   photo-electric 190-1, 204, 272
   single photo-electric 190, 271
   single photoelectric 294, 296, 309
   third 233
motive, psychological x, 5, 7-9, 315-16, 320
mystery 123, 132, 233, 319, 322, 324, 335, 337, 349-50
mysticism iv, vi, 5, 15-21, 24, 30-1, 128, 130, 165, 244, 295, 333-7
   modern 19-20

N

net magnetic field 163
neutrinos 222, 233-8
neutrons 128, 135-6, 175, 186, 190, 208, 213, 219, 221-2, 234, 237-8, 361-2
   extra 135-6
Newton 12-13, 28, 121, 146, 170, 183, 185-6, 208, 316, 353, 356
Nikola Tesla 241, 247, 262-3
number, atomic 134-5

O

objective God, personal 327-8
objective magnetic field 39, 189
objective reality 3-5, 14, 18-21, 24, 31, 33, 37, 39, 67-8, 71-2, 125-6, 149, 189-90, 227, 229-30, 341-3
   observing 339
objectivism xiv, 29-30

Objectivity in Matter vi, 125-6, 128, 130, 132, 134, 136, 138, 140, 142, 144, 362
oblivion 15-16, 19-22, 26, 32, 229, 236, 244, 263, 297, 312, 315-16, 319, 327, 330
observation ix, xiii-xiv, 3-7, 10-14, 26-8, 31-8, 41-5, 65-8, 70-1, 73-6, 91, 101-2, 111-12, 114-15, 127-8, 317-19
   act of 2, 5, 14, 41, 43-4, 110, 144, 157
   apparent xvii, 10, 14, 23, 277
observational cause 335, 337
observational space-time reality 33
observer, sideline 37, 73
Olbers' Paradox 279, 281, 292, 307, 309, 311
orbit, photoelectric particle 292
original meanings ix-x
oscillating magnetic field 256
outer shell electrons 266, 271

P

Pagels 185, 363
particle density 139, 159, 197
particle fields xvi, 190
   electric xv, 139, 189, 265
particle light 97, 139, 249, 256, 287
particle model 299
particle nature 192, 197, 243
Particle of Intelligence vii, 226
particle physics 62, 236-7, 349
particle theory 93
particles
   electric 117, 143, 164, 182, 189, 242-3, 255
   ghost 233, 235
   illusive 349-50
   smallest 140
   wave verses 291

Index | 372

pervading magnetic field   209
philosophy   ii, ix-xii, xiv, xviii, 6, 9, 18, 22, 25-31, 33, 161, 189, 227, 232, 329-31, 333-5
photo-electric field   172, 186-7, 189, 191-6, 200, 202-3, 209, 223, 226, 246, 251, 266, 283, 355
photo-electric monopole field   283
photo-electric particle field   140, 143
photo-electric particles   138-9, 150, 191, 199-200, 209, 233, 262, 283, 285, 332
photo-electric quantum field   193
  single   193
photo-electric spiral   191, 202, 287
photoelectric effect   xvii, 150, 161, 324, 326
photoelectric particles   122, 141, 154, 175, 292, 296, 308
Photographic Time   82
photon bundle   139-41, 159, 175, 294, 296
photon helix   156, 159
photon particles   89-90, 172, 293
photon spirals   141, 152, 156, 158, 172, 175, 191, 284, 287, 308
photons
  electric field of   267, 270
  very low energy   178
  visible   216, 263
photons spiral   268, 303
physical reality   17, 52, 101, 103, 117, 168, 186, 188, 243, 283, 315, 336-7, 342, 349, 359
physical space   68
Plato   xi, 3, 27, 346
polarity   202-3, 206, 209, 212, 222
polarization   183, 208, 210, 214, 218-21, 273, 288-90, 299-301, 309
polarized light   288, 300-1, 303
polarizing filter   300-2, 306

positive ions   179
positron   225-6, 349
Predicate Nature of Energy   vii, 145-6, 148, 150, 152, 154, 156, 158, 160, 362
predicate reality   xiii-xiv, 10, 19, 23, 33, 35, 46, 130, 342
predicative   xv, 2, 4-6, 9-10, 14, 16-17, 29, 31, 44, 52, 128, 148, 329
probability   208, 280, 296, 300, 330, 342-3
propagation, plane of   199, 283-4, 308
properties, photo-electric   201, 251
proportion   vi, 41, 47-50, 54, 56-7, 64-6, 126, 133-4, 165-8, 172, 227-8, 231, 308
  space/time   52, 55, 126, 130
psychology   ii, xix, 31, 231-2, 243, 319-20, 330, 334, 336

Q

quantum electrodynamics   189-90
quantum particles   vii, 156, 308
quantum physics   339, 355-6
quantum relativist   150, 341
quantum theory, subjective interpretation of modern   325
quarks   186, 208, 222, 236, 277

R

radar   75-6, 153
radiation, electromagnetic   237, 241, 243
radio   xvii, 263, 282, 286
radio waves   vii, xvii, 241, 245-9, 252, 254-7, 260, 262, 272, 275, 285-8
red shift   109, 279-81, 287-8, 291-2, 295-7, 300, 303, 306, 308, 331
reflected light   90, 288, 304-6, 308, 310
reflection   6, 218, 256, 298, 301, 303, 305-9

refract 118, 198, 200, 246, 256-7, 280-1, 283-4, 286, 303, 305, 307

refraction 98, 120, 142, 151, 156, 159, 193, 199, 207-8, 280-2, 284-5, 287, 291, 300, 306-9

relative meanings 5-6, 9

relativity
  general 39, 60, 104, 150-1, 310, 325, 354, 356, 362
  theory of 18-19, 32-3, 96, 116, 182

relativity and common sense 361-2

relativity and epistemology vii, 161-2, 164, 166, 168, 170, 172, 174, 176, 178, 180, 182, 184, 362

relativity equations 31, 52, 104, 118-19, 150, 183, 268
  general 324

relativity physics 62-3, 128, 177

religion i-ii, 5-6, 18-21, 25, 31, 128, 130, 232, 244, 313, 321, 334, 336, 345

responsibility i, ix, xi, 16, 21, 23, 53, 130, 167, 177, 239, 311, 319-23, 334, 336, 360

rest energy 63, 141, 147, 172, 174, 176

right action 232, 297, 331-2, 342, 360

rod, moving 86-90, 100, 112-13, 119

S

scalar quantity 57, 60, 126, 170

silicon 210, 212-14

singularity 357-9

slits 158-9, 247, 300-1, 303, 306

Smith, Joseph 346-7

Socrates xi, 27, 30, 319, 346-7

sound, velocity of 79-80, 83-5, 92

space
  conceptual 66-7

conservation of 18, 321

curved 13, 67, 126-7, 131-2, 151, 263, 268, 353

distort 33-4

objective 140

relativist curving 339

three-dimensional 33-4

space curvature 127, 151, 262-3
  matter causes 151

space curves 32, 229

space/time 49-50, 56, 59, 66, 101, 126, 129, 152-3

space-time 151

space/time, observer's 101-2

space/time components 59

space/time relation 55, 73, 130

Space/Time Vertical 72

Special Relativity 5, 37, 90, 101, 110, 163, 295, 324

Special Relativity Principle 29, 31, 38, 59, 106, 108-9, 112, 164, 179, 183, 203, 288, 324, 343

Special Relativity Theory 97, 105, 107-9

Special Theory xv, 20, 86, 96, 109-10, 281, 285, 326

Special Theory of relativity 71, 76

Special Theory of Relativity xviii, 33, 85, 91, 94, 96-7, 100, 103, 119, 155-6, 160, 183, 292, 295, 318, 325

spectrum 6, 286, 290-1, 309, 364
  visible 272, 304-6, 309

spiraling photons 165, 172, 225, 348

Spirit of God 329, 331, 342

spot meter 302

Square of Reason vi, 41-2, 44, 46, 48, 50, 52, 54, 56, 58, 60, 62, 64, 361

string theory vii-viii, 356-9

subjective reason   xii, xiv, xviii, 3, 7-8, 37, 230
subjective space   68, 171, 341
subjectivism   xiv, 26, 29
sun's magnetic field   152, 224
sunset   280-1, 307
symmetry   7, 46, 52

T

Tao   vii, 334, 338, 340, 342
tetherball   181-2
tetherball players   182-3
theory, general   xv, xviii, 111, 119, 121, 162, 325-6, 352
theory updates   vii, 345-6, 348, 350, 352, 354, 356, 358, 360
time, absolute   116, 341
transformation equations   36-7, 44-5, 114
   classical   91, 100
   proper   80, 86
transformer   194, 196-7, 214
Tree of Life   230-2
TV antenna   259, 261, 282
twin paradox   109-11

U

ultraviolet light   139, 280-1, 304
understanding   viii, 7, 9-13, 26, 34-5, 45, 58, 127-31, 139-40, 171-2, 203, 231, 243-4, 311-14, 332-7, 348-50
universal laws of existence   232
universe
   expanding   ix, 296, 325-6
   objective   71, 112

V

vector   50, 56-7, 73, 78, 119, 170, 173, 251
velocity
   negative   84, 88, 91, 93, 98, 100, 114-15, 119, 164
   negative directional   80, 83, 87
   positive   87, 98, 100, 110, 114
velocity of light   87, 90, 100, 113-14, 193
velocity reversal   110
verb   xiv, 3-4, 6, 11, 20, 35, 37-8, 43, 46, 53-4, 62, 66, 69-70, 149, 218, 231
volt photons   202, 204, 216
voltage   195-200, 202, 204, 213-14, 217, 257, 269-72

W

warped space   ix, 14, 34, 64, 153-4, 156, 233, 326
watts   195-6, 198-9
wave   xvii, 59, 105-6, 120, 147-8, 156-7, 242-3, 245, 247-9, 252-7, 281-3, 285-6, 291, 294, 299-300, 334
   electromagnetic   242-3, 273
   television   xvii
wave fields   188
wave mechanics   99, 107, 157, 159
wave propagation   99, 108, 247
wave theory   93-4, 121
   classical   92
wavelength   107, 198-9, 247, 258-60, 281
wisdom   viii, 7, 231, 277, 313-14, 318, 339
work   xiv, xvi, 14, 57-8, 68-9, 79, 91, 96, 116, 119, 126, 230-1, 240-1, 316-19, 346-7, 349-50
world   xiv, 17, 163, 227, 232, 318-19, 323, 325, 327, 331, 345

Y

Young, Arthur M.   79, 361-2

Z

zero point   37, 51-2, 55, 73-8, 80, 83-91, 93, 100-1, 110, 115,

119, 129, 164, 259

zero point of measurement
36-7, 73-5, 80, 129

If you have read the Einstein Illusion and would like to contact Samuel Dael to discuss any subject in his writing, please feel free to do so, but be prepared to make sense:

sd@samuel-dael.com

The Einstein Illusion

www.ingramcontent.com/pod-product-compliance
Lightning Source LLC
Chambersburg PA
CBHW021759220426
43662CB00006B/114